101 CASE STUDIES IN CONSTRUCTION MANAGEMENT

This book provides 101 real-life construction management case studies from an author with over 40 years' experience in the construction industry and as a lecturer in construction management. Over 14 chapters, Len Holm has included case studies from real jobsites that cover organization, procurement, estimating, scheduling, subcontractors, communications, quality and cost control, change orders, claims and disputes, safety, and close-outs. Other hot topics covered include BIM, sustainability, and lean.

Each case is written in straightforward language and designed to test the reader's independent and critical thinking skills to develop their real-world problem-solving ability. The cases are open to interpretation, and students will need to develop their own opinions of what's presented to them in order to reach a satisfactory solution.

The cases are ideal for use in the classroom or flipped classroom, for individual or group exercises, and to encourage research, writing, and presenting skills in all manner of applied construction management situations. Such a broad and useful selection of case studies cannot be found anywhere else. While there is often no "right" answer, the author has provided model solutions to instructors through the online eResource.

Len Holm is a Senior Lecturer at the University of Washington, USA. He has over 40 years' construction industry experience at all levels and owns his own construction management firm. He is the author of numerous books on construction, including *Management of Construction Projects: A Constructor's Perspective*, 2nd edition, also published by Routledge.

101 CASE STUDIES IN CONSTRUCTION MANAGEMENT

Len Holm

Routledge
Taylor & Francis Group

LONDON AND NEW YORK

First published 2019
by Routledge
2 Park Square, Milton Park, Abingdon, Oxon OX14 4RN

and by Routledge
711 Third Avenue, New York, NY 10017

Routledge is an imprint of the Taylor & Francis Group, an informa business

© 2019 Len Holm

The right of Len Holm to be identified as author of this work has been asserted by him accordance with sections 77 and 78 of the Copyright, Designs and Patents Act 1988.

British Library Cataloguing-in-Publication Data
A catalogue record for this book is available from the British Library

Library of Congress Cataloging-in-Publication Data
Names: Holm, Len, author.
Title: 101 case studies in construction management / Len Holm.
  Other titles: One hundred and one studies in construction
  management | One hundred one studies in construction
  management
Description: Abingdon, Oxon : Routledge, 2019.
Identifiers: LCCN 2018020482| ISBN 9780815361978 (hardback :
  alk. paper) | ISBN 9780815361985 (pbk. : alk. paper) | ISBN
  9781351113632 (ebook)
Subjects: LCSH: Building—Superintendence—Case studies. |
  Construction industry—Management—Case studies.
Classification: LCC TH438 .H65 2019 | DDC 690.068—dc23
LC record available at https://lccn.loc.gov/2018020482

ISBN: 978-0-8153-6197-8 (hbk)
ISBN: 978-0-8153-6198-5 (pbk)
ISBN: 978-1-351-11363-2 (ebk)

Typeset in Bembo
by Swales & Willis Ltd, Exeter, UK

Visit the eResources: www.routledge.com/9780815361985

CONTENTS

The 101 case studies are generally organized according to their primary topic or "chapter," as indicated below. The order of the 14 chapters follows the textbook *Management of Construction Projects: A Constructor's Perspective*, 2nd ed., Routledge, 2017, by John E. Schaufelberger and Len Holm.

There are several case studies that fit within each of the above primary topics. Each of the 14 chapters begins with a general topic overview and its own detailed table of contents. Many cases relate to other sections as well (most of the examples cross many topics). A complete detailed table of contents with all 101 cases is included with the instructor's manual (see www.routledge.com/9780815361985).

ILLUSTRATIONS

Figures

Tables

ABOUT THE AUTHOR

Len Holm grew up in a construction family. His father Arne Holm was a high-end residential and light commercial general contractor in Grays Harbor County, Washington. Len was shoveling concrete and driving nails from the age of 10. He eventually became a journeyman carpenter and foreman. Len was the only member of his family to go to college, and earned bachelor's degrees in Building Construction and Economics and a master's degree in Construction Management, all from the University of Washington. Len's first job out of college in the early 1980s was as an estimator and a scheduler traveling around the country building power plants for one of the largest construction firms in the world. He later found his way back to Seattle and worked on numerous high-technology, medical, and industrial projects with a large local general contractor as a project manager, senior project manager, and company stockholder. Len founded his own company, Holm Construction Services, in 1994, which has provided a variety of construction consulting services on hundreds of residential and commercial projects, including owner's representation, expert witness, and contractor training. He has been an instructor at the University of Washington since 1993 and has taught over 110 quarter-long courses on 13 different topics to over 3,500 students. He has authored and co-authored several books and articles on a variety of construction issues, including project management, estimating, and dispute resolution. Three books are standard textbooks for many construction management programs throughout the United States and in other countries: *Management of Construction Projects: A Constructor's Perspective*, 2nd ed., 2017, Routledge, and *Construction Cost Estimating: Process and Practices*, 2005, Pearson, both co-authored with Dean John Schaufelberger, Ph.D., and *Introduction to Construction Project Engineering*, 2018, Routledge, co-authored with Dr. Giovanni Migliaccio. Another new book is in construction close-out with Routledge as well, with an expected release in late 2018, titled *Cost Accounting and Financial Management for Construction Project Managers*. Questions or comments regarding this case study book may be sent to holmcon@aol.com. I hope you enjoy the stories.

ABOUT THE ILLUSTRATOR

Barbara Holm is a stand-up comedian, writer, author, artist, and actor based out of Portland, Oregon. She has performed in several comedy festivals across the United States, and was recently seen on NBC's *Last Comic Standing* TV series, as well as *Laughs TV* and *Portlandia*. Barbara has written for several publications, including IGA and the Huffington Post. She was awarded Time Out New York's Joke of the Week and was named "one of the best things about comedy" in 2012. Her comedy has been described as clever, unique, idiosyncratic, and exuberant. Barbara has published several graphic novels, which are also available on Amazon. Follow Barbara's work on Twitter @barbara_holm or look her up at barbaraholm.com.

PREFACE

This book began many years ago with a collection of just a few short stories. There was a desire within the Department of Construction Management at the University of Washington to strengthen the students' written and verbal communication skills. The stories transformed into case studies, and were first used to improve the students' presentation skills within a project management course and also to provide practice for the subsequent quarter's capstone course. At that time, the department was just becoming active in student competitions. The activities of research, working in a team environment, preparing both a written and verbal response, and competing among fellow students helped the department raise their competition teams to national recognition.

A few construction stories were added each year. This book now includes 101 case studies, with subsections for several cases and numerous questions for each one. The sources of the cases are mostly from projects that I have personally been involved with, from over 40 years in the construction industry as a carpenter, project engineer, project manager, owner's representative, construction consultant, and expert legal witness. A few of the topics have been donated from other friends in the industry. The cases included here represent concepts from actual construction projects, but no real construction project has been used in this book, and any similarity with an actual project is coincidental. Several industry professionals and professors at the University of Washington have been using and supplementing to this body of work for two decades.

The first edition of this book was titled the same as this new edition, *101 Case Studies in Construction Management*. It was a self-published manuscript and available only for my students and local contractors, which used it as an in-house training tool. The second and third editions were retitled *Who Done It? 101 Case Studies in Construction Management*, and were made available to other universities through Amazon/CreateSpace. This new format has returned to its original title, but now includes a few pages of construction management topic introductions and a couple of figures with each chapter.

As indicated, the initial reason for compilation of this material into one book was for the benefit of all students and the faculty at the University of Washington. Now the material is utilized within several construction management university programs and also for in-house contractor training across the United States and internationally. There are many potential uses for this material, including a writing course, competition practice, verbal presentation skills, testing, an applied case study, a capstone course, and/or an upper-division or master's-level research course.

This material has been successfully used to interact with the students after they have read their standard project management course textbook and listened to a lecture on any one subject. For example, the students are to read about claims before class. They hear a short lecture, which reinforces their project management textbook. Maybe after a break, either again in teams or individually, they now present responses and debate a case or two that applies to that specific topic, in this case claims.

An instructor or corporate facilitator could easily add a single line to any of these cases to customize the material for their course or company or desired outcome. With some individual creativity from an instructor or construction firm facilitator, the possibilities are unlimited. Several questions are included for each case study. Many others are possible, which could allow this material to expand exponentially. Contractors apply many of these examples to their own projects. "Lessons learned" exercises presented during in-house corporate training sessions or professional seminars are very beneficial to "outside of the box" continued learning.

The first self-published project management book by this author was titled *The Project Manager's Toolbox*. This was also a popular resource for many construction companies for their in-house training programs. Quips were included throughout the text and coined "toolbox quotes." These quotes have survived that original manuscript and have now been incorporated in *101 Case Studies in Construction Management* and noted in *italics*, as well as a few figures and graphics utilized from lectures.

The organization of this book follows that from our very successful project management textbook *Management of Construction Projects: A Constructor's Perspective*, by John Schaufelberger and Len Holm, also published by Routledge. Both of these books follow the natural course of a construction project, from start-up through close-out. A few advanced topics are included at the end of this new format for the ambitious professor and student. This new edition also includes chapter introductions and topic overviews, which allow it to also function as a stand-alone introduction to project management course or an advanced master's-level course structured around the use of case studies and group discussion.

Construction is a people business. Any contractor with a business license can place concrete or install structural steel, but it is people that build construction projects. Construction companies and construction projects have a multitude of different potential arrangements. Chapter 1 highlights some successful and some not so successful construction organizations. Project owners procure contractors either through bid or negotiated processes, and several case studies are included in Chapter 2 that analyze procurement processes. Chapter 3 has many case studies that emphasize the importance of good construction contracts, including insurance and bonding issues.

A good estimate is the foundation for any construction project, just as a stable concrete foundation is important to any building. General contractors rely on subcontractors for 80–90% of their work, and receipt and analysis of subcontractor bids are among some of the case studies in estimating in Chapter 4. Chapter 5 transitions from estimating to scheduling and schedule control – which is a major responsibility of the contractor's superintendent. Because subcontractors account for so much of the work on any construction project, proper management of subcontractors and suppliers can be the difference between a successful and unsuccessful project for the general contractor and the project owner. It is a challenge for this former project manager not to fill up the entire case study book with subcontractor stories, but Chapter 6 has been limited to just 11 interesting examples. But as is the situation with any of these case studies, the subject matter often crosses into several other chapters and subject areas, so the reader will find subcontractor discussions throughout the book.

Once the project owner has chosen the general contractor and the general contractor has chosen their subcontractor team, it is time to start the project. Chapter 7 on start-up also has case studies on preconstruction services, mobilization, and value engineering. We as construction managers do not use hammers and saws as do carpenters; written communication tools such as requests for information (RFIs) and submittals are in our toolbox. People build buildings, and people do not always draft clear construction contracts or communicate clearly. Chapter 8 discusses some good and bad examples of contractor communications. One of the most important things any contractor does is get paid. Chapter 9 utilizes several case study examples of the importance of an efficient pay request process, including lien and retention issues. One of the other important construction management aspects necessary for contractors and project managers to be successful is cost control, as the stories in Chapter 10 emphasize. Case studies for an advanced topic of lean construction have also been added to the cost control chapter for this reformatted edition.

Even if the project is brought in on time and within budget, it will not be successful if the quality is unacceptable or if anyone was hurt. The other two important pillars of construction management success, quality and safety controls are the subjects of Chapter 11. All projects have change orders, and it is the proper management of the change order process, including change order proposals and formal contract changes, to which the jobsite team must also focus their attention. Change order case studies are included in Chapter 12, and claims – those change orders that are not readily approved – are included in Chapter 13, along with dispute resolution techniques. The final chapter is a mix of other new and advanced topics, such as contract close-out, building information modeling

(BIM), and sustainability. Abbreviations are used throughout the construction industry. Learning construction abbreviations and acronyms is similar to learning a foreign language, and this book has liberally shared that language with the reader. All of the abbreviations utilized within this work have been referenced in Appendix A, and a glossary of construction management words and terms used in this book and in the industry is included in Appendix B.

At the request of several professors, an instructor's manual was developed that includes "potential" answers to some of the cases and questions – there are not necessarily any exact answers, although some are more correct than others. This manual has been aptly titled *Suspects*, and is available to instructors through the publisher (see www.routledge.com/9780815361985). This edition includes several new cases on hot topics such as lean construction, BIM, and sustainability. A few previous cases were rolled off and some renumbering has occurred to fit the new ones in. A comparison of old case numbers to new case numbers is also included with the instructor's manual if the reader has the benefit of an older edition and wants to find where their favorite case study is now located.

I would like to thank my good friends and adjunct professors Mike Matter, Sara Angus, and Chris Angus who used the first rendition of this book in several of their courses. I will be forever grateful to the now thousands of students who have enjoyed the book with me and repeatedly remarked how it enhanced their learning experiences. If you have ideas of new cases for future editions, or comments or suggestions, please send to Len Holm at holmcon@aol.com. I hope you enjoy the stories.

1

CONSTRUCTION ORGANIZATIONS

Including personnel issues

Introduction

There are three primary organizations or companies involved in all construction projects: the project owner, the architect, and the general contractor (GC). On heavy-civil projects, the civil engineer or structural engineer will be the primary designer and possibly an architect will work for them. Several other design consultants are involved as well, including mechanical and electrical engineers, landscape architects, and others. There are many ways these companies can be organized, including a traditional GC organization, agency construction manager (CM) or construction manager at risk, and design-build (DB) organizations. Figure 1.1 is a simple traditional organization chart with the three primary participants represented. Other more advanced arrangements, but still including

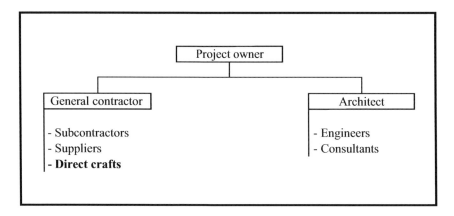

FIGURE 1.1   Traditional organization chart

these three primary parties, are design–build–operate–maintain (DBOM), integrated project delivery (IPD), and public–private partnership (PPP).

Every contractor structures the size and makeup of a specific project management organization or team depending on the size of the project, its complexity, and its location with respect to other projects or the contractor's home office and contract influence. The two major participants in any jobsite organization include the project manager and superintendent. Even subcontractors have these same two participants, although they may have different titles such as account executive and foreman. The cost of the jobsite project management organization is considered project overhead, and must be kept economical to ensure the cost of the contractor's construction operation is competitive with other contractors. The goal in developing a jobsite project management organization is to create the minimum organization needed to manage the project effectively. If the project is unusually complex, it may require more technical people than would be required for a simpler project. If the project is located near other projects or the contractor's home office, technical personnel can be shared among projects or backup support can be provided from the home office. If the project is located far from other contractor activities, it must be self-sufficient.

General contractors employ subcontractors to perform 80–90% of the work and perform some carpenter and laborer scopes in-house, such as concrete, wood framing and trim, doors, door frames, and door hardware. Some GCs will also employ ironworkers to install rebar in concrete and erect structural and miscellaneous steel.

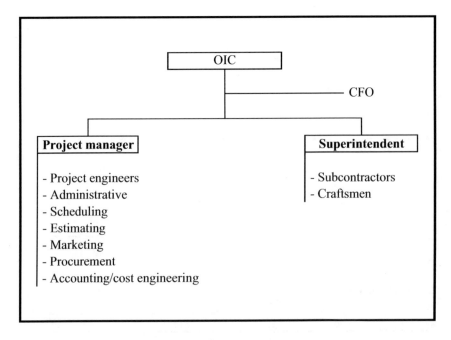

FIGURE 1.2 GC's jobsite organization chart

The choice of project management organizational structure depends on the contractor's approach to managing projects. A simple GC's jobsite organization chart is shown in Figure 1.2. Some project owners will seek alternative methods to contract for design and construction services, including employing design engineers and construction subcontractors directly. The case studies in this chapter highlight some of the risks of these alternative organizations.

Case studies

Case studies in this chapter include:

1. Pass-through contractor
2. Owner's contracts
3. Construction manager or general contractor?
4. Design-build joint ventures
5. Bait and switch
6. Multiple contracts
7. Ambitious project manager
8. Client expertise
9. Developer or general contractor?
10. Overworked project engineer
11. Self-performed failure

Most of these case studies overlap with other primary topics. Case studies 46, 47, 53, 61–64, 88, and 89 also involve organizations.

Toolbox quote

*Just like construction management and construction manager,
the terms project management and project manager can mean different
things to different people.*

Case 1: Pass-through contractor

A commercial construction firm that reports to be a design-build general contractor actually operates more as a pure construction manager. They sell their services as being an agent for the owner. This firm will procure all design and construction services, originate the contractual agreements, charge a fee on all hard and soft costs, but have the second-tier firms execute their contracts with the project owner directly. This method of procurement is referred to as a *pass-through* contract. The CM does not sign or initial any of the contract documents, pay requests, change orders, requests for information (RFIs), submittals, etc. The CM explains to owners the attractiveness of this contracting arrangement as a means of saving additional tax and insurance markups. The CM does not have problems convincing the subcontracting and supplying firms of the advantage of contracting directly with an owner, as these firms are now one step closer to the owner, and subsequently the bank. Several of this CM's projects have had problems with second-tier contractors (such as foundation settlement, lack of sufficient air conditioning, window leakage, roof leakage, or dead landscaping). In each instance, the CM has stepped back, not protecting the owner, and forces the owner to resolve problems with subcontractors directly. On other projects, there have been problems with the owner not paying the contractors or designers, and again the CM has stepped back, not protecting the subcontractors. The second-tier firms have had to pursue resolution, through liens or other means of collection, with the owner directly. The pass-through CM receives a guaranteed lump sum fee for their services and jobsite general conditions are cost-reimbursable. What is wrong with this system? How can it be improved and still utilize the services of a construction management or owner's representative firm or individual? How would a general contractor or CM at risk system improve this condition, if at all? Develop three organization charts (pass-through, owner's rep, and CM at risk) and utilize them to answer these questions.

_____

_____

_____

_____

_____

Case 2: Owner's contracts

a. On a very large, very complex remodeling and expansion project, an owner has contracted with numerous members of the design team independent of the architect. The architect does not have any second-tier design firms under their contract. The owner has also contracted with many different material suppliers and specialty subcontractors directly and not through the general contractor (GC). The owner is of the belief that they can save money avoiding multiple markups by doing business this way. The owner has, however, written into every contract that each firm is still responsible to "coordinate and communicate directly with the other independent designers and contractors as if they were contracted under the standard vertical method," rather than this horizontal method. What risks has the owner assumed? What risks are the designers and contractors assuming through this system? Draw up organization charts depicting this condition and the "standard" condition.

_____

_____

_____

_____

_____

b. Assume the following costs for the case described above. What would the cost increases have been to the owner if they had contracted in a normal fashion? Does this also provide *contracting opportunities* for the contractors or designers? Explain why or why not.

- $50 million total construction cost.
- $30 million total construction cost of sub-consultant-designed work.
- Four percent total architectural fee on the architectural portion of the work, and they would receive an additional 10% markup on sub-consultants' design fee if run through the architect's books.
- The sub-consultants would receive 10% design fee of the construction cost of their portion of the design, either billed directly to the client or through the architect.
- Five percent general contractor fee on costs run through their books.
- The values for four sample subcontracting firms working directly for the owner are $550,000 for landscaping, $950,000 for a swimming pool vendor, $5 million for the electrical subcontractor, and $12 million for the mechanical subcontractor. Assume the subcontractor fees are included in these figures. The other subs traditionally work through the GC.

_____

_____

_____

_____

_____

Case 3: Construction manager or general contractor?

A client is ready to negotiate a contract with a construction firm for a $30 million shelled office building project. The design-development documents (DDs) are complete. The building permit has been applied for and is scheduled to be issued in two months. The architect has requested the owner now bring on a contractor to assist with the balance of preconstruction services, estimating, scheduling, constructability analysis, material selections, and value engineering (VE) during the construction document (CD) development phase. The client and the architect have received written proposals and conducted interviews, and have narrowed the short list down to two firms who have a completely different approach to contracting. Both appear to be equally qualified with respect to experience, references, availability, etc. Both firms have worked with the architect and the owner successfully on previous projects. Both firms are quoting a competitive 4% fee on top of the cost of the work. All other conditions are equal. The only difference between the two firms is that one is a pure construction manager and will subcontract 100% of the project except jobsite administration. The other is a typical general contractor. The GC is only interested in building the project if they are allowed to perform the work that they customarily self-perform, such as concrete, carpentry, reinforcement steel, structural steel, and miscellaneous specialty installation, which will account for 30% of the cost of the work on this shell.

a. Take the position of the CM. Why is it to the owner's and the architect's advantage to employ your firm during the preconstruction phase? What are the advantages of using a CM during the construction phase? Discuss all project control issues, including quality, safety, schedule, and cost. Is a CM at risk truly an owner's representative? Why is it better for you that your firm is selected now and on board when the tenant improvement projects become available? Does the GC hide costs? Sell your position and be creative. Use the tools you have learned from this book, your classes, professional experience, and outside research to convince the owner that the CM procurement approach is more advantageous than the typical general contractor.

_____

_____

_____

_____

_____

b. Take the position of the GC. Cover the same issues as outlined in the CM position above, except with the opposite perspective. Does the GC have more control over cost, schedule, quality, and safety? Which firm can build it better, faster, and safer? Who is best at looking out for the client's interests?

_____

_____

_____

_____

_____

Case 4: Design-build joint ventures

A client negotiated a design–build contract with a joint venture (JV) general contractor and architect for the design and construction of a $100 million chemical manufacturing facility. The project was constructed within a reasonable time frame and for a fair price. The quality of the work was acceptable and there were not any time–loss safety incidents. The architect's and the contractor's joint venture was dissolved immediately after substantial completion was achieved and they had received their retention. In only six months after completion, the owner had achieved 90% of potential production output and sales were higher than anticipated. About that time, several employees began feeling ill and several citizens living nearby and working outside the facility filed a claim due to unacceptable odors coming from the plant. The government shut the plant down. It was discovered that there was a major design error in the mechanical exhaust system and it would cost the owner $10 million to redesign and repair, and maybe 10 times this in lost revenue. There would eventually be personnel claims as well due to illness. The owner took the designer to arbitration and won. The owner separately took the contractor to arbitration and won on the basis that the general "should have known" that the design was in error. Should the owner be allowed to pursue the two parties separately or only the JV? Was the general at fault? Shouldn't the owner have had some responsibility in this? The city approved the plans – what sort of liability do they have? What types of insurance will come into play here? Does this mean the design-build procurement approach is flawed? What sorts of checks and balances are required of the design for a design-build project?

_____

_____

_____

_____

_____

Case 5: Bait and switch

This contractor conducted considerable research before submitting their negotiated proposal to the owner and architect teams. The GC proposed management and supervisory individuals that were both familiar to the reviewers and had relevant project and location experience. The market was very busy. After award and contract execution, the GC systematically changed out all four members of the team who were proposed and participated during preconstruction. Each change was explained and apologized for. Can they do this contractually? Is this ethical? Is this "bait and switch" customary on negotiated projects? Who loses when new team members are brought on board? When might it be to either the contractor's or the owner's advantage to bring on new individual team members?

_____

_____

_____

_____

_____

Case 6: Multiple contracts

Some owners will contract with multiple general contractors on the same site. This scenario is sometimes referred to as _multiple primes_ or _five primes_. Some owners will also employ specialty subcontractors directly. What sort of risks and coordination issues is an owner assuming by not placing all of the work on one site under one GC? Does this save or cost the owner money? Do these sorts of multiple contracts also place risks on the contractors or designers? Does this also provide for "contracting opportunities" for the contractors? When might this not only be an acceptable delivery method for the project owner, but a preferred one?

_____
_____
_____
_____
_____

Case 7: Ambitious project manager

This relatively new yet very ambitious general construction project manager (PM) was given an assignment for a repeat industrial client. The project was bid lump sum at $50 million by a staff estimator. It was the largest lump sum project ever undertaken by the GC at that time. There was a $1,000,000 bid error. The officer in charge (OIC) directed the PM to get every change order possible. The PM was not to worry about the possibility of future work with the client. During the one year of construction, the PM lost his project engineer and was not allowed to replace him with any experienced help. He had to hire from outside. He also did not receive any home office supervisory help. The OIC would meet him for lunch once a month at a remote site. Why did the OIC distance himself? Why assign this project to a new PM? The PM worked seven days a week – sacrificing health and family time to bring the project back into the black. The project was ultimately brought in with a clear fee of $600,000, but relations had been damaged; the GC did not get to work with the client again. Later in the PM's career with other contractors, he was not welcomed back on this client's site. The PM did what he was asked to do, didn't he? He was successful, wasn't he? Did the PM make any errors? What would you have done?

_____
_____
_____
_____
_____

Case 8: Client expertise

a.   During the course of construction of a 500–unit apartment complex, the developer has hired the project manager/estimator (PM1) away from the project's general contractor and also hired the architectural project manager (Arch1) away from the prime designer. Both of these people were involved in the project during preconstruction. Why would the developer do this? What complications could arise from this for all parties? Is this ethical?

_____
_____
_____

budget. The owner had a bad prior experience with a fully designed-bid-built mechanical project. Under the recommendation of the architect and estimating consultant, they decide to employ design–build (D-B) mechanical and electrical subcontractors directly. Do you agree utilization of D-B MEP subcontractors helps control costs? What role could a third-party MEP designer play with respect to "criteria documents"? How much are design fees for MEP subcontractors? Who should carry these design contracts, the owner, the architect, or the general contractor? Who carries the errors and omissions (E&O) insurance for this work? Who is named as additional insured? What risks do all of the parties assume?

_____

_____

_____

_____

_____

b. Assume in the above case the client carries the design portion of the MEP contracts and a negotiated general contractor will ultimately carry the construction portion of the contracts when a guaranteed maximum price (GMP) contract is developed. The design agreements are very loose and the subcontractors' deliverables and detailed schedules of values (SOVs) are not defined. The sub-contractor-design team reports that the documents are complete and the general contractor submits their GMP to the owner. It is significantly over their earlier budget, especially in the areas of mechanical and electrical. The subcontractors refuse to negotiate their construction estimate and the owner announces they will bid this portion of the work out. Is this ethical? Is the owner guaranteed a lower construction price if this portion of the work is put on the street for open bidding? Will the original design–build subcontractors submit competitive lump sum bids on their own drawings? If so, will they be successful submitting change orders for discrepant documentation? What are the risks for all parties?

_____

_____

_____

_____

_____

c. The design contracts are silent with respect to ownership of the design documents. The subcontractor-designers claim they own the documents. Because the owner has paid 90% of the design fee, with the final 10% invoice in process, the owner also claims it owns the documents. Who is correct? What would standard contract verbiage dictate?

_____

_____

_____

---

d. A third-party engineering firm and two subcontractor competitors review the drawings, and agree that they are maybe only 50% complete and are not biddable. All of the details and equipment schedules are missing. The specifications are in outline form. The owner requests that the original contracted MEP subcontractor-designers finish the drawings. The MEP firms indicate that the owner has received what they paid for and that it is customary for the final details to be completed through the shop drawing and request for information (RFI) processes. Are they right? If the documents are bid at this level, will the industry "fill in the blanks"? If they are bid, who will take ownership of the design? How is this issue resolved? If the client does not pay the remainder of the 10% due on the design contracts, can the design-build subcontractors lien the property even though they have not provided any physical material improvements to the site?

Case 13: Bid or negotiate?

Two contractors are pursuing the same client for a dental clinic. The drawings are now 50% complete. The dentists' organization knows and trusts both contractors. The first contractor likes the lump sum market and is encouraging the client to complete the drawings and competitively bid the project. The second contractor is recommending that the owner allow a select group of contractors to prepare proposals utilizing a negotiated procurement and GMP pricing approach based on the drawings as they exist now and save additional architectural fees.

a. Argue the first contractor's case. What are the advantages to the owner to bid the project? Use statistics and materials obtained from your courses, textbooks, and outside research to support your position.

---

b. Argue the second contractor's case. What are the advantages to the owner to negotiate the project? Use statistics and materials obtained from your courses, textbooks, and outside research to support your position.

_____
_____
_____
_____
_____

Case 14: CM/GC accounting

This public works project is being built under the construction manager/general contractor (CM/GC) or CM at risk procurement method. This is also known by some as GC/CM. The CM/GC negotiates an early maximum allowable construction cost (MACC) with the client. This is similar to a GMP. The construction manager (CM) was also awarded the concrete and steel packages (work they normally self-perform as a GC) under separate competitive lump sum bids. The CM is now a subcontractor to itself.

a. How can the client be sure that costs incurred under the lump sum portions of the project by the CM's subcontractor team are not being charged to the CM's MACC? How are the CM and subcontractor kept at arm's length with respect to cost and contractual issues? Does the CM evaluate the subcontractor's change orders and pay requests critically? Does the CM help the subcontractor with site supervision and equipment usage more than they may with other subcontractors? Is this procurement system fair to the taxpayers? Are the CM's MACC accounts auditable? Are the CM's accounts auditable for their own lump sum subcontracted portions of the work?

_____
_____
_____
_____
_____

b. One reason many public agencies are choosing the negotiated CM/GC MACC approach over traditional competitive bidding with general contractors is to reduce claims and lawsuits. Can these results be validated? CM/GCs offer constructability reviews, value engineering, and early procurement of long-lead materials through this delivery method as well. List some of the other advantages. Can the exact payback or return on investment be determined for the client from these services? If you were assigned the task of researching and "proving" these claims, what methodology would you choose?

_____
_____
_____
_____
_____

Case 15: Bid at 50%

A semiconductor manufacturer has decided to build a new satellite facility. This project will eventually cost $200 million. It is their corporate policy to bid all work. Before a significant amount of design can be developed (soft costs spent), their board of directors must give approval on a maximum anticipated cost. This is a catch-22 situation as it is difficult to develop a maximum cost until significant design is available. The owner lacks in-house capability to develop their own estimates. At 50% design completion, the owner solicits "bids" from four general contractors.

a.   As a general contractor who has experience in this market, should you pursue bidding this project? Why or why not? If the designer is a reputable firm with whom you have a good team relationship, does this change your decision? What alternate procurement methods should the owner consider?

   _____

   _____

   _____

   _____

   _____

b.   The architect on this project recommends to the owner that they should use design-build subcontractors for the mechanical and electrical portions of the project early in the design process. Is this a good idea? Why would the architect suggest this? Should the owner follow this recommendation? If so, what sort of guidance or criteria would you recommend the owner follow in the selection process of these two important subcontractors? When should these firms be brought onto the team? What selection process should be followed: low fee, résumé, construction estimate, design approach, others? As the general contractor, if the D-B MEP subcontractors are already on board, does this change your decision to pursue this project? Does the general contractor assume more or less risk with design-build subcontractors?

   _____

   _____

   _____

   _____

   _____

c.   The semiconductor client decides to save additional design fees and competitively bid out the design-build subcontractors concurrent with the general contractors. Is this a good idea from the owner's perspective? Is this fair to either the subcontractors or the general contractors? How can the general contractor bids now be evaluated properly? Who carries the design portion of the contract and the associated errors and omissions (E&O) insurance for D-B subcontractors?

_____
_____
_____
_____
_____

d. As the general contractor pursing this project, your firm decides to team
    up with a design–build mechanical and plumbing subcontractor during the
    bid process; this is termed a strategic alliance or strategic partnership. You
    work together closely to assure that your scopes of work are consistent. Your
    mechanical subcontractor's price is $80 million. On bid day, you receive an
    unsolicited proposal from an outside mechanical firm that was teaming with
    one of your general contractor competitors. This subcontractor is a very com-
    petent firm that has experience in this industry as well. Their proposal appears
    complete. Their price is $75 million. What do you do?

    _____
    _____
    _____
    _____
    _____

e. List other methods this owner may have chosen to get valuable competent
    early MEP and GC estimates. Can any of these methods "guarantee" a final
    cost to the owner?

    _____
    _____
    _____
    _____
    _____

Case 16: Union or open shop?

You are presenting your company to a client who has a 50,000 square foot
$10 million tilt–up concrete construction project to award. The project as planned
is now a "shell" with potential tenant improvement (TI) work to be negotiated
at a later date. The client has utilized both union and open shop (or merit shop)
construction firms on prior projects. The client does not have a bias toward either
choice of labor scenarios. The client has narrowed their contractor choice down
to two relatively equal firms. Both contractors have extensive résumés in this type
of work. Both contractors have successfully worked with this client previously.
Which is the best system of labor, union or open shop? One team (a) is to assume
the union position and argue their point; the other team (b) is assigned the open
shop position. Try to anticipate the points your competitor will make when preparing your own presentation. Convince the client that your firm has the best

choice of labor. Use information presented in your textbooks and classes, as well as outside research, to form the basis for your presentation. Cover subjects such as quality, schedule, flexibility, cost, training, and safety. Assume that the project is not being built in a labor-bias area, such as the northwest and northeast are predominantly union whereas the southeast is predominately open shop. Be creative. Only one team will be awarded the project. The other team will get nothing.

_____

_____

_____

_____

_____

Case 17: Executive home

a. A pharmacist and his wife have agreed to a purchase-sale agreement with a sole proprietor contractor/developer for an executive yet speculative $5 million waterfront home. The contractor had not yet started construction when the agreement was signed. $200,000 is put into escrow for earnest money. All remaining funds are to be transferred upon closing, which will occur after certificate of occupancy (C of O). The buyers were provided with six pages of plans and an artist's rendering from which their agreement was based. There were not any separate specifications. The house is scheduled for a 12-month construction duration. What risks is the contractor taking? What risks are the buyers taking?

_____

_____

_____

_____

_____

b. In a custom home scenario, the customer employs the contractor to build their "dream home." In a speculative home situation, the buyer receives what the contractor intends to supply. In this case, the pharmacist and his wife expect custom features and upgrades, considering the purchase price. The contractor is quick to point out at every turn that either: (a) the drawings said plastic laminate, not marble countertops; or (b) the 40 cents per square foot allowance for paint over drywall does not cover cherry wood wall paneling. The contractor requests change orders for all of the upgrades and the buyers agree. The purchase-sale agreement requires all change orders to be paid up front in cash. Is this a standard clause? Why does the contractor require this? Over the course of construction, almost $1 million will be added in change orders and paid up front by the buyers before closing. Now what risks do the two parties have?

_____

_____

_____

_____

_____

c. During the construction process, the residential construction market has improved dramatically. All of the new high-technology millionaires are driving up the sales prices of these executive homes. Approximately halfway through, the builder realizes he could have sold this speculative home for $7 million without the buyers' upgrades. With the upgrades (which have all been nice improvements, including a dock and wine cellar), the house is worth well over $8 million. The builder is regretting he put the home on the market so soon. The builder acts as his own project manager. He also has a full-time on-site superintendent. Both of them do everything they can to make the process miserable for the buyers. Construction on this site noticeably slows down. They claim the busy market is taking away their subcontractors and quality craftsmen. They know the buyers have sold their previous home and are uncomfortable in temporary quarters. The buyers are really pressuring the contractor to speed up. The only thing they speed up is the original contract work that is certain to require rework due to contemplated upgrades by the buyers. This drives up the cost of the changes even further and adds to the buyers' frustration level. Why is the contractor causing these conflicts and delays? What recourse does the buyer have? If the buyer walks away from the purchase-sale agreement, who keeps the earnest money? Who keeps the $1 million in change orders?

_____

_____

_____

_____

_____

Case 18: Successful schools?

a. This school district had a long record of working with repeat general contractors. Even though the state law required competitive bidding, the same general contractor would usually be the low bidder. Quite often the second-tier subcontractors were also the same from project to project. The architect and sub-consultants were also consistent on every project despite state requirements for competitive proposals. The contractor, architect, and the school district all thought quite highly of each other. After 10 projects together, none of the parties had ever claimed each other. When their success was audited by an outside consultant, he found that the client carried a 15% contingency after the bids were received. Is this a reasonable amount considering that 100% plans and specifications were prepared? At the end of the project, almost none of the contingency was left. What was happening here? Were the taxpayers being treated fairly? Were other competing contractors and designers being treated fairly? Isn't our ultimate goal a quality project without claims?

_____

_____

_____

_____

_____

b. Many professionals in the construction industry feel designers are currently preparing less-than-complete documents today. The competition for projects has in many cases caused design fees to be lowered. This same school district asked their auditor to review their architect's fees and see if additional fees would result in better documents. As stated above, the school district almost exclusively used the same design team. In response to the question, the architect indicated that they thought their fee was already fair. The state law determines that the architect's fee is approximately 7% and is not negotiable by either party. The law, however, does not discuss limits on change orders. What the auditor discovered was that the designer had actually been receiving a total fee of almost 10% due to design service change orders upon project completion. They were paid additional design fees to remedy discrepant documents. Was this fair? Is everyone better off by just letting this issue pass? How should design fees or designer selections be made on public projects? How are they made for private projects? Do you feel that lower design fees relate directly to poorer-quality design? Can this be substantiated one way or another through research?

_____

_____

_____

_____

_____

Case 19: Public set-asides

The voters of this state have recently approved a new issue that abolishes preferential treatment, or "set-asides," to special interest groups. Explain what set-asides are. Explain the terms MBE/WBE/SBE/VBE/DBE. What are the state-mandated percentages of work that needed to be awarded to these groups on public projects such as a high school in your location? How many dollars would these percentages represent on a $100 million project? How will this state's new legislation affect these types of firms? How will this new legislation affect the way these construction projects are bid, awarded, and constructed? What is the current status of this legislation in your state? Why are some voters now overturning these old laws? What will be the outcome? Take either the pro set-aside position (a) and argue against this legislation, or the open markets position (b) and argue against state-imposed special interest percentages. Use information from your classes, textbooks, and outside research.

_____
_____
_____
_____
_____

Case 20: Public alternatives

a.   Many public entities such as the federal government, public universities, and
      cities and states are evaluating alternative procurement methods in an attempt
      to avoid the negative claim atmosphere that currently surrounds lump sum bid-
      ding. What are these alternative methods? Who favors each method (public,
      government, large contractors, small contractors, design community, minority
      contractors)? Do public bid jobs end up with more claims? Substantiate your
      position with facts from projects that utilized alternative procurement methods.
      What is going right with this movement? What is wrong? What can you rec-
      ommend to solve these problems? Base all of your conclusions on information
      you have learned in this book, your classes, and outside research. Be creative.

_____
_____
_____
_____
_____

b.   Argue in favor of the *traditional open market lump sum bid approach*. Is this truly a
      "bid 'em and sue 'em scenario"? Why is the public perception incorrect? Use
      facts and figures to support your position.

_____
_____
_____
_____
_____

c.   Argue in favor of *short-listing qualified subcontractors and general contractors*. How
      can short lists be fairly prepared? Can this reduction of competition improve
      relations, quality, safety, cost, and schedule control? Does this restrict compe-
      tition and drive up prices to clients? Which is most fair to the taxpayers?

_____
_____
_____
_____
_____

d.   Prepare an argument in favor of the *CM/GC alternative procurement method*. Is this
      fair for all firms, including the taxpayer? Does the CM/GC have an incentive to

understaff the project? Will other traditional general contractors bid the structural subcontractor package to the CM/GC? Would the CM/GC treat other GCs fairly? If the CM/GC gets awarded the structural work as a lump sum subcontractor to themselves, how is the accounting of actual construction costs kept separate?

_____

_____

_____

_____

_____

e. Prepare an argument in favor of the *design-build delivery method*. Can this work? Is it working? Who provides the checks and balances to assure the successful firm is not "cheapening" the design to save on construction costs? Who approves the design documents? Who approves submittals? Who responds to RFIs?

_____

_____

_____

_____

_____

f. *Integrated project delivery (IPD)* is a fairly new delivery method where the GC, owner, and architect all sign one agreement and equally share in the risks. How often has this been done, and does it work?

_____

_____

_____

_____

_____

g. *Public–private partnership (PPP)* is another method to sidestep competitive bidding. A private owner will design-build-operate-maintain (DBOM) a facility (or even a road or bridge) for a public client/tenant such as a university, in exchange for market rents guaranteed far into the future, such as 20 years. At the expiration of the lease, ownership may transfer to the tenant. How is this fair to the taxpayers? How does a developer acquire a PPP project?

_____

_____

_____

_____

_____

h. Describe formal *partnering*. Does mandatory partnering affect your position on any of the above delivery methods? Will it solve some of the problems? How much does partnering cost, and which party pays for it? How much does it save?

_____

_____

_____

_____

_____

Case 21: Labor agreement

A unionized public water district solicited bids for the construction of a $500 million water reservoir project. The request for quotations required all bidders to accept a "project labor agreement," which was included in the contract documents and negotiated between the water district and the local unions. This was a complex project, and the water district was concerned about the qualifications from potential out-of-town open shop construction firms. Bids were received from several "union" contractors, with the lowest at $490 million and one open shop contractor at $480 million. The open shop bidder was disqualified based upon the above requirements. The water district awarded the bid to the low union firm on the basis that they complied with the bid documents and their bid was within the project budget. The open shop contractor took the water district to court on the basis that the "project labor agreement" violates state law, which requires open and competitive bidding, and the owner is required to hire the low bidder. Was the disqualification fair to the open shop firm? Is this fair to the taxpayers? Will they be successful in overturning this award? If the $480 million firm wins the award, will the $490 million firm have a basis for their own claim? How do government agencies generally address these situations in their bid documents? What are Davis–Bacon wages or prevailing wages? What if this scenario was with a private utility provider?

_____

_____

_____

_____

_____

Case 22: Negotiated success

The client, architect, and professional owner's representative team all agreed that bringing a negotiated general contractor on board early during the design process was the best procurement method for this project. Five contractors proposed on this private build-to-suit corporate office building. The proposals included:

- individual team member résumés and similar corporate projects;
- safety records;
- financial references;
- current and anticipated volumes;

- milestone schedule; and
- preliminary budget.

a. The budgets were based upon early schematic documents. One contractor's budget was $18 million. Three were in the $22 million to $25 million range. The fifth was high at $30 million. The contractor chosen had an average length schedule and a budget of $23 million. Why was the selection made for this construction team member? Why was the contractor with the low estimate not chosen? Why do you think their budget was so low? Why was the contractor with the high estimate not chosen? Why do you think their budget was so high? Why did the selection team require a budget to be included with the proposal at this early stage? Do you think budgets should have any bearing on the construction team member selection?

_____

_____

_____

_____

_____

b. The above office building project was extremely successful in many regards, but like most projects, it had its ups and downs. The general contractor's original budget was slightly more than the client had anticipated. A six-week value engineering (VE) process ran in parallel to the architect's preparation of the design–development documents. Some VE items were accepted and incorporated. The owner also ran a new pro forma based upon other market conditions and increased the budget. After completion of the construction documents and negotiation of a $22.5 million guaranteed maximum price contract, construction commenced. The following summarizes some of the project's statistics:

- There were 100 change order proposals (COPs). Some of these were for credits.
- There were 250 requests for information.
- The project finished three months behind the originally proposed schedule.
- There were problems with two major subcontractors, as discussed below.
- There were not any outstanding liens or claims at close-out.
- The final contract amount was incredibly close to the originally proposed budget.
- The contractor did not make their full fee, but still realized a fair profit.
- There were not any savings to share.
- All of the team members are looking forward to working together again.

Should every project conduct a VE process, even if it is within budget? Why do you suppose the project was considered so successful? How could the budget and the final costs have been so close?

_____

_____

_____

_____

_____

c. Assume the opposite result. The client was very dissatisfied. Using materials presented above and below in this case, prepare an owner's claim, highlighting the project's "failures" rather than successes. Feel free to use some creativity.

_____

_____

_____

_____

_____

d. One of the reasons this project ran late was because it had a slow start with a shoring subcontractor. The general contractor competitively bid all of the normal subcontract areas, including shoring. They chose the low bidder, who was a large but out-of-state specialty contractor. This firm imported their people and equipment and also engaged several other second-tier subcontractors for portions of the work. Early in this phase of the project, the owner's representative witnessed what appeared to be a very poorly managed operation. He mentioned this to the contractor's site superintendent. Was he right to do this? Should a client or designer become involved in a contractor's "means and methods"? What sort of liability issues might arise?

_____

_____

_____

_____

_____

e. The superintendent's response to the owner's representative was that he was staying out of the subcontractor's management problems. Was the superintendent right to do so? Could he contractually force the subcontractor to change foremen, equipment, or second-tier subcontractors?

_____

_____

_____

_____

_____

f.   The shoring subcontractor ultimately finished one month late. The general contractor was never able to recover from this delay. The subcontractor also submitted several unsubstantiated COPs six months after demobilization, all of which were rejected. What are the lessons learned here for all parties?

_____

_____

_____

_____

_____

g.   The other problem subcontractor on this project was responsible for the window wall system. Review the following chronology:

- Three individuals from an experienced glazing firm recently began their own firm and were the successful bidder for this project.
- The subcontractor was not able to post a performance and payment bond, but the GC pledged to watch them closely.
- The subcontractor subcontracted (brokered) all of the work to third-tier firms.
- Submittals were turned in late.
- They submitted on an alternate manufacturer. After extensive discussion of potential schedule impacts to use the specified manufacturer, and verbal guarantees from the GC, approval was given for the alternate.
- The submittals were incomplete and required several revisions.
- The subcontractor relied on a third-party company to verify field dimensions.
- Materials arrived from another third-tier fabricator late.
- The original third-tier installation subcontractor was changed out at the last minute.
- The new installation subcontractor could not adequately staff the project.
- The GC and the architect immediately questioned the quality of the installation.
- Lack of proper submittals and accurate field dimensions failed to identify a discrepancy in the original construction documents. Whose fault is this?
- The manufacturer, fabricator, and installer were not being paid timely by the prime glazing subcontractor/broker; therefore, the GC was writing two-party checks.

What errors were made by the project team, which caused or failed to prevent these issues? Drawing organization charts may assist with understanding some of these complicated cases with inherent communication problems.

_____

_____

_____

_____

_____

h.   The general contractor stepped up and assumed direct control of the window wall package. The first-time prime subcontractor was paid off and their contract was properly terminated. The GC then contracted directly with the manufacturer, fabricator, and installer. The direct management of these third– (now second–) tier firms by the GC was now much more intense. This portion of the project was finished with acceptable quality and within days of the original duration. Was the GC therefore successful in this area? If so, it was largely due to the GC's PM, who had many tools in his toolbox, as illustrated in Figure 2.2. What sort of risks did they assume when taking control of the work? Did they protect the client? Who ultimately provides the guarantee of the windows?

_____

_____

_____

_____

_____

_____

FIGURE 2.2   Project manager's toolbox

3

CONSTRUCTION CONTRACTS

Including insurance and bond issues

Introduction

A construction contract is a legal document that describes the rights and responsibilities of the contracting parties (for this discussion, particularly the owner and the general contractor). The terms and conditions of their relationship are defined solely within the contract documents. These documents should be read and completely understood by the contractor before deciding to pursue a project. As shown in several of the case studies in this chapter, understanding contractual agreements is something that is not always adequately handled. The contract is also the basis for determining a project budget and schedule. To manage a project successfully, the project manager must understand the organization of the contract documents and understand the contractual requirements for his or her project. Knowledge of the terms and conditions and workings of the contract is essential if a project manager expects to satisfy his or her contractual requirements.

The prime contract describes the completed project and the contractual relationship between the owner and the contractor. Usually, there is no description of the sequence of work or the means and methods to be used by the contractor in completing the project. The contractor is expected to have the professional expertise required to understand the contract documents and select appropriate subcontractors or qualified tradespeople, materials, and equipment to complete the project safely and achieve the specified quality requirements. For example, the contract documents will specify the dimensions and bolted or welded connection requirements for structural steel, but will not provide the design for temporary shoring, safety, or hoisting; these are of the contractor's choosing.

Contract documents

Contracts are either standard or specifically prepared documents. Standard contracts generally are preferred because they have been legally tested in and out of the legal system. Such documents have been developed by several professional organizations such as the American Institute of Architects (AIA) and ConsensusDocs. A typical construction contract consists of five primary documents, as exhibited in Figure 3.1. All of these documents must work together in a complimentary fashion:

- contract agreement;
- special or supplemental conditions;
- general conditions;
- technical specifications; and
- drawings.

Any of the contract documents can be modified by addenda or by change orders. Addenda are issued by the owner prior to award of the contract, and change orders are executed between the owner and the contractor after contract award. Any addenda, or later change orders, are also considered to be contract documents, assuming they are listed in the prime contract agreement.

The *prime agreement* describes the project to be constructed, the pricing method to be used and cost, the time allowed for construction, and the names and points of contact of the project owner and the contractor. Supporting documents may be incorporated into the agreement by reference or as exhibits,

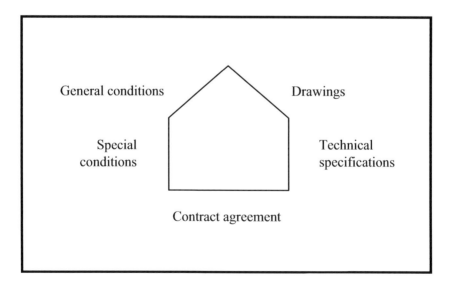

FIGURE 3.1 Contract documents

some of which are listed here. These documents then also become part of the contract documents.

Some project owners use a *project manual*, which contains the *special conditions*, *general conditions*, soils report, erosion control requirements, prevailing wage rates, *technical specifications*, and other project-related documents. The invitation to bid (ITB) and request for proposal (RFP) are generally not considered contract documents unless referred to in the prime agreement or included in the project manual.

Special or *supplemental conditions* are customized for each specific project. They include issues such as work times, liquidated damages, and site and parking conditions. Special conditions may be included in the project manual or part of Construction Specifications Institute (CSI) divisions 00 or 01 within the technical specifications book.

The *general conditions* provide a set of operating procedures that the owner typically uses on all projects. They describe the relationship between the owner and the contractor, the authorities of the owner's representatives or agents, and the terms of the contract. Some owners use standard general conditions published by professional organizations such as the American Institute of Architects' AIA A201.

The *technical specifications* provide the qualitative requirements for construction materials, equipment to be installed, and workmanship. The old 16 CSI divisions have been replaced with a new listing of 49 divisions, but construction managers will find examples of each being utilized today. The *contract drawings* show the quantitative requirements for the project and how the various components go together to form the completed project. The drawings represent final placement and configuration of construction materials and systems, but not how the work is to be accomplished.

Contractors and project owners use a variety of risk management strategies, including insurance and bonds. The requirements for these will also be included in the prime contract agreement. A couple of case studies involving insurance and bonds have been included in this chapter as well.

Contract development and execution

There are many steps involved in creation and execution of the prime contract agreement, too many to cover in this case study book. Suffice it to say that there is more to contract drafting, negotiating, and execution than one party simply filling out a generic template and the other signing it. As some of these cases represent, it appears that the agreement is not always fully understood by both parties. In order for a contract to be considered executed, there needs to be an offer to perform services, an acceptance of that offer, some conveyance or transfer from one party to the other, both parties have to be authorized to enter into an agreement, and

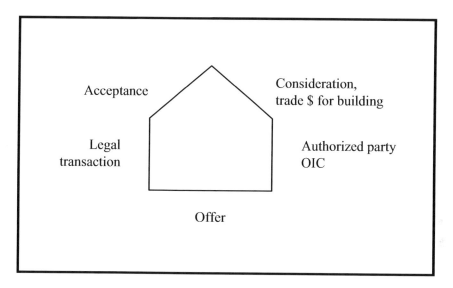

FIGURE 3.2 Contract house

the transaction must be legal. The contract house, as shown in Figure 3.2, reflects these five requirements.

Case studies

Case studies in this chapter include:

23. Moving target
24. Budget or bid?
25. Turnkey impasse
26. Historic restoration
27. Residential dispute
28. Shared savings
29. Subcontractor bonds
30. All-inclusive?
31. Seismic repairs
32. Line item estimate
33. Allowance accounting
34. Contract concerns

Most of these case studies overlap with other primary topics. There are many cases that also involve contracts, including 3, 4, 5, 11, 12, 14, 55, 57, 72, 79, 83, 96, and 97.

Toolbox quotes

Just like good fences make good neighbors, good contracts make good construction projects, and good construction contractors.

All construction documents are important, but the most important construction document is the contract.

Case 23: Moving target

This large $400 million public works project was competitively bid utilizing drawings that everyone thought were relatively complete but turned out were just a progress set. The contract format was to utilize a guaranteed maximum price (GMP) rather than lump sum, which would have been more customary for bid and public works projects. A fee of 4% was stated in the GMP. Shortly after the contractor started the project, a completely new document set was issued. The contractor was to continue with construction based upon these new documents and prepare a change order, or revised GMP, in parallel. One month was allowed for re-estimating. Almost as the estimate was being finished, a third completely new document set was issued. The client put aside the requirement for change order number one since this third set of documents was really the set that was going to be built to, and hopefully incorporated via change order. The contractor then began to estimate these documents. This same process repeated monthly throughout the two years while the project was under construction. A total of 20 complete sets of revised documents were issued. The contractor continued on with the project and kept very accurate accounting records. The quality of the work, schedule adherence, and safety awareness were all managed well. The contractor eventually turned in a very large change order reflecting an increase of almost 50% over the original "bid," backed up with "actual" costs, not estimated costs. Aren't actuals more accurate than estimates? Was this a lump sum project? Was it a conventional GMP project? Or was it actually completed as a cost-plus-percentage-fee project? Was this fair to the taxpayers? Many public projects end up with the contractor claiming the owner. Do you think that happened in this case?

Case 24: Budget or bid?

A residential client hired a general contractor (GC) to construct a major $3 million executive home. The client contracted separately with a reputable architect for the design. After the permit was obtained, the architect's contract was closed out. The contract authored by the contractor discussed in detail what was reimbursable but did not tie the GC to a fixed price. The $3 million estimate is referred to as a "budget" in the contract. The owner had assumed because of verbal discussions with the contractor prior to the contract execution, and because of early estimates provided by the GC on paper where the word "bid" was used, that this was a lump sum $3 million agreement. These discussions and estimates were not tied to or referenced in the contract. There was not a third-party owner's representative or agency construction manager (CM) involved, and the owner had never been involved in a construction project. During the course of construction, the owner and the city had requested several changes, but none of them were formalized into the contract. Many of these were additive but some were also deductive changes. Many of the changes were due to building code changes. Because the general contractor understood this to be a time and materials (T&M) project, they had not felt it was necessary to submit change orders against a "budgeted" amount. The GC had previously invoiced and received 67% of the original budget, or $2 million, from the owner. These invoices were all reviewed and approved by the lender. At approximately 90% completion, the GC invoiced the owner for the full 100% amount of the original $3 million estimate. When pressed, the GC indicated that the project will overrun the budget by approximately $500,000. The owner and the GC disagree. The owner refuses to approve the current pay request. Can he do that? The GC pulls off the job and refuses to do any more work until the owner agrees to the revised budget of $3.5 million and pays the $1 million now due. Can the contractor legally do that? The owner dismisses the GC at this point. Both parties sue each other. How did this happen? Who is at fault? What should have been done to prevent this situation? What do they do now?

_____

_____

_____

_____

_____

Case 25: Turnkey impasse

a. A privately held small industrial manufacturer has come to an impasse with their general contractor. The client is a first-time builder and engaged the contractor to design, permit, and build a relatively simple $5 million pre-engineered steel building. There was neither an architect nor a professional owner's representative on the project. Most of the space is to be open-span with a bridge crane running full length. The crane will be contracted separately directly from the manufacturer to the client. The GC planned bare-bones offices and restrooms in a small attached structure. The dispute is in regard to the level of finishes and a mezzanine in this attached structure. The GC's chief executive officer (CEO) owned the local franchise on this type of design–build facility. He provided the drawings and obtained the permit, and therefore sees himself as the interpreter of the documents. Is he correct? The GC has refused to perform this questionable scope unless they receive a change order for $50,000. Can an owner force a contractor to proceed with work that is in dispute? The client stopped the current progress payment of $200,000 for other work already completed. Can they do this? The contractor subsequently stopped all work and demobilized. Can they do this? Is this a good negotiating tactic?

_____

_____

_____

_____

_____

b. A mediator was brought on board. When the mediator asked to review the contract, the client indicated that the contractor had not provided one. The contractor responded that there was a contract, and produced the original two–page proposal that was signed "received" by the client. The client had not been provided with an original or copy of this proposal after signing. Without any further formal agreements, is this signed proposal the contract? The conditions of this proposal were consistent with the contractor's actions. When is a turnkey contract beneficial to each of the parties?

_____

_____

_____

_____

_____

c. Take the position of the contractor and make your five-point case to the mediator.

_____

_____

_____

_____

_____

d. Take the position of the industrial manufacturer and make your five-point case to the mediator.

_____

_____

_____
_____
_____

e. Take the position of the mediator. What recommendations will you make to each of the two parties?

_____
_____
_____
_____
_____

Case 26: Historic restoration

The following facts are available for the dispute resolution parties to evaluate:

- This is a historic restoration (substantially different from a remodel or renovation) project. The home is over 100 years old and is listed on the national historical register. The owner (and also occupant during renovation) is a medical doctor.
- The owner has employed a one-person architectural office. The architect is licensed and she is reported to have this type of restoration experience. She is a single mother with a toddler. She has her child with her at most meetings with the owner, contractor, suppliers, and the city.
- The architect recommends the owner engage a general contractor who is reported to have completed similar projects in San Francisco. There is not any documented résumé or brochure from the contractor in evidence that supports either that he made these representations or that he has this experience.
- The owner of the construction firm also works as a carpenter, superintendent, and project manager on his projects. His annual volume is approximately $200,000. This was his sole project for the two years he works on it. He invoices his time as "cost."
- The architect provides a one-page list of work that "may" need to be performed during the restoration project. There are not any drawings or specifications. The list uses many terms such as "if needed," "as necessary," and "if requested by the owner." Foundation repair is not one of the items listed. Most of the items on the list relate to finishes such as millwork, light fixtures, and wallpaper. This list is not referenced or attached to the contract.
- The GC provides the owner a one-page budget of $700,000 for the work. The owner reports that this is over his budget, and they ultimately agree on a reduced figure of $600,000. The one-page budget is marked up in hand with the revisions to achieve this figure. Neither party initials the revisions. The revised budget is not dated and also not attached to the contract. All of the figures on the budget are round figures. The contractor reportedly did not obtain outside subcontractor or supplier input or perform any detailed cost estimating quantity take-off.

- The GC authors a contract to the owner. This is a "homegrown" document. There are not any general conditions. Neither the budget figure nor the budget document, nor a construction schedule, are referenced or attached to the agreement. The contract specifies the hourly wages to be paid to the craftsmen. It turns out later that the actual wages paid to the craftsmen are less than those specified in the contract, although the contractor bills according to the contract. The contract specifies that the GC will be paid 15% overhead and 10% profit on top of the cost of the work. Terms such as "overhead," "profit," "reimbursable," or "cost" are not defined. Neither the owner nor the contractor obtained legal assistance to review the contract.

- The contract is silent with respect to change issues, schedule, dispute resolution, subcontract practices, and retention. The contract requires the GC to "bid out subcontract work," although subcontract is not defined.

- The contractor begins the project without any drawings or specifications or a building permit. None of this is finalized during the entire two years the contractor is employed. The contractor obtains several minor "subject-to-field-inspection" permits for small portions of the work.

- The architect only produces rough 8½×11 hand-drawn sketches, all without reference numbers or issue dates. There are not any directional documents such as construction change directives (CCDs) or construction change authorizations (CCAs) provided. The best documentation available (after deposition) from the architect is her personal log. The bulk of the documentation later available from the contractor are photographs of actual site conditions and construction progress.

- The owner continues to make changes and add scope to the work during construction, such as an underground wine cellar, a swimming pool, and an underground garage. Many substantial floor plan changes are made. All of the existing plumbing and electrical and insulation is determined to be inadequate and requires replacement. Significant foundation and roof structure damage is discovered and requires extensive rework.

- Many unknown conditions are discovered, such as solid bedrock under the foundation, which must be jackhammered and removed by hand. The existing concrete foundation system has failed and lacks any reinforcement steel, and must be replaced. The entire four-story house must be jacked up to accommodate this work. The sanitary sewer pipes are completely plugged, and some of the craftspeople are infected with giardia from working around raw sewage.

- The contractor employs many subcontractors during the process, authorizing all of them to proceed on a T&M basis and signing the subcontractors' proposals rather than issuing subcontract agreements. This also applies to suppliers and their purchase orders. The subcontractors later make these documents available to the court. The general contractor evidently did not retain copies. The doctor claims the GC was contractually required to obtain lump sum subcontractor bid quotes.

- The owner contracts with all the second-tier design firms, such as civil and structural engineers, directly. Many of these firms are rotated in and out of the project. There are not any second-tier design firms that start on the project who also finish the project.

- The owner sometimes dictates to the contractor which subcontractors to use, but the subcontractors' money is run through the GC's books, all with markups. The owner will, from time to time, fire a subcontractor, also without going through the GC. The owner will communicate directly with the subcontractors on change of scope issues. Very few of these directions are documented, except the occasional reverse memorandum from a subcontractor.

- There were not any written requests for change orders originated or executed by any of the three parties during the two years that the general contractor was engaged.

- The contractor invoices the owner every two weeks and is paid promptly until the last pay request.

- The owner replaces the first architect 16 months into the contractor's two-year employment. The second architect is also a one-person shop, but he documents to the other extreme. On an average day, he will write 5 to 10 formal letters to many of the parties involved. Many of these letters use very direct and often adversarial language.

- During this second architect's tenure on the project, which lasts for only six months, he continually complains in writing about the lack of definitive direction from the owner and the lack of as-built field conditions or available materials from the contractor. The lack of this information impacts his ability to finalize the drawings.

- During the first 24 months, there were not any formal liens filed by any of the designers, contractors, or suppliers. In fact, there is not any record of material men's notices or conditional lien releases in the files.

- The original general contractor is dismissed after two years of employment. He has reportedly spent $1.2 million. He has completed approximately 10% of the work that eventually will be required to finish the home according to the doctor's intentions. Many of the interior finish items on the first architect's scope list were not even started by the original GC during his tenure. The GC does not receive payment on his last $60,000 invoice. The GC liens the project.

- Several months after dismissing the first GC, the doctor files a lawsuit. He claims the GC had an obligation to perform all of the necessary work and achieve the original budget. He claims the GC misreported his accounting data. The doctor claims the contractor did not have a right to charge his own wages to the job, as he was the owner of the construction firm. The GC also charged a 2% bookkeeping fee for his wife's accounting efforts for accounts payable and accounts receivable. The doctor does not feel the wife's bookkeeping charges are cost-reimbursable. The doctor claims that some of the work accomplished does not meet code, and other work was of such poor quality that it required rework by subsequent contractors. The owner's claim is for $400,000.

- During deposition, none of the original three primary parties are able to produce any written documentation such as memos, diaries, letters, phone logs, change orders, meeting notes, requests for clarifications, submittals, etc.
- The owner will eventually go through two more GCs and one more architect. As of this writing, more than four years after the project was started, it is still under construction.

a. Argue the contractor's case to the arbitration panel using at least three project management tools. Use facts from both within and outside of this case and your classes. One side generally wins and one side loses in arbitration. Be creative. Anticipate the owner's response in your preparation. You only get one chance to present your case. You will not get a rebuttal.

_____
_____
_____
_____
_____

b. Argue the doctor's case to the arbitration panel using at least three project management tools. You only get one chance to present your case. Use facts from both within and outside of this case and your classes. Be creative.

_____
_____
_____
_____
_____

c. As a third-party arbitration panel, what would your judgment/decision be? How much is the award, and to whom? Take a firm position. Base your decision on the information you have learned in your textbook, classes, these presentations, and the "documents." Be right. Sell the arbitration method to the class. Why is arbitration better than mediation for this specific case?

_____
_____
_____
_____
_____

d. Take the position of a court of law. Assume that the arbitration process above was "non-binding" and the losing party appeals the case to you. Do you uphold or overturn the arbitration panel's findings? Substantiate your decision. Why do you feel a trial court is better than arbitration, mediation, or a dispute resolution board (DRB) for a case such as this?

_____
_____
_____
_____
_____

Case 27: Residential dispute

The following facts have been presented by two opposing parties for the dispute resolution teams to decide upon:

- This is a dispute between a general contractor and a client on a custom home project.
- The owner owned the land prior to the agreement.
- The GC authored the contract, which was a homegrown document (i.e. non-copyrighted).
- The GC borrowed some of the AIA A201 general conditions items, but not all, did not incorporate nor reference the AIA document, and modified those that he did copy.
- There wasn't a third-party owner's representation from a CM firm, nor did the architect provide these services.
- The architect was from out of town and did not have any prior relationship to either party. The architect's involvement was discontinued at the time of issuance of the building permit.
- The contract does not indicate it was "cost-plus," but this was the builder's intent.
- The contract does not indicate it was "lump sum," but this was the client's intent.
- The project was negotiated between the two parties. There were not any other competitive bids received.
- There was only one change made to the original agreement by either party, and it was inserted by the owner and agreed to by the contractor. It indicated that "The only items which are not in the contractor's control which can result in changes to the contract amount are increases in the purchase cost of materials and scope increases originated by the owner. These items, if exceeding the contract amount, will be performed on a time and material basis."
- The original estimate by the contractor was $650,000. This was value-engineered (VE) down to a mutually agreed contract amount of $580,000.
- The value engineering changes included shelling some of the interior space, reducing the square footage of the house from 5,300 square feet (SF) to 4,400 SF, and (verbally, according to the GC) reducing some of the finish and appliance estimates. None of the VE items were recorded on paper or incorporated into the contract.
- The contract references the original permitted and estimated drawings.
- There are not any finish or appliance specifications in any of the documents.

- During the course of the project, the client's wife would choose finishes and appliances from the builder's recommended suppliers.
- According to the builder, he advised the client's wife that they were over-spending the estimate. None of this is documented. There were not any signed change orders to the original agreement.
- There is not any discussion in the documents or in the contract regarding allowances for finishes and appliances.
- The builder finished the house and has invoiced the client for approximately 10% over the contract amount for a total of $640,000. The client has paid up to the contract amount of $580,000.
- The city issued a certificate of occupancy (C of O) and the client has moved in.
- The contractor filed a lien on the home for $60,000. This lien has prohibited the owner from obtaining permanent financing. The owner is still paying on his construction loan, which is now 2% above current market rates. The bank is pressuring the owner to obtain permanent financing.
- The GC paid all of the suppliers and subcontractors in full, and none of them filed liens.
- The GC has essentially been paid its "cost," but has not received any fee, which coincidently amounts to approximately 10%.
- The GC and the owner had worked successfully together on other professional endeavors. Both have good reputations in town. The client is an attorney by trade.
- The contract mandates arbitration as the dispute resolution method.
- The owner has filed a counter-suit due to damages associated with the lien and financing for $30,000.
- Both parties are also attempting to recover legal fees.

a. Argue the contractor's case to the arbitration panel using at least three project management tools. Use facts from both within and outside of your classes. One side will win and one side will lose in arbitration. Be creative. Anticipate the owner's response in your preparation. You only get one chance to present your case. You will not get a rebuttal.

b. Opposite to the above, argue the owner's case to the arbitration panel using at least three project management tools. You only get one chance to present your case. Use facts from both within and outside of your classes. Be creative.

c.   Where is the architect in all this? Take the position of the architect, as the inter-
     preter of the contract documents, and propose a resolution to this situation.

_____

_____

_____

_____

_____

Case 29: Subcontractor bonds

Your firm has negotiated a 30-story high-rise office complex in a downtown
metropolitan area. This is a very major project both for your firm's volume and
reputation. Unfortunately for you, as the project manager, and your client, the
market was very busy when the subcontracts were bid. Subcontractors with whom
you do not have prior relationships or history will perform many of the major
scope categories. Your guaranteed maximum price proposal and contract with
your client requires that your firm post a 100% performance and payment bond.
The cost of this bond was anticipated and is included. At the time of the execution
of the bond, your bonding agency is requiring that you also bond all second-tier
subcontractors and suppliers whose values are greater than $40,000. This is a total
of approximately $15 million worth of subcontracts. Their average bond price is
2%; therefore, the subcontract bonds will cost approximately $300,000. This value
was not included in your GMP estimate. You approach the client and ask them
to pick up these bond fees, but they respectfully decline. Should these firms be
bonded? Is it standard that your bonding agency would require these bonds? Is the
client required to pay? Where did you error? What can you do now?

_____

_____

_____

_____

_____

Case 30: All-inclusive?

A developer and an architect bid out an $8 million mixed-use development
(MXD) project to three general contractors. The drawings are less than perfect,
maybe 60% complete at best. One reason for bidding at this stage was the decision
by the developer to hold down the architect's fee. None of the contractors asked
any questions during the bid cycle. The documents include many phrases such as:

*   Build according to industry standards.
*   Must meet all codes and city requirements.
*   The documents are not meant to show every detail.

- The contractor is responsible to report discrepancies prior to contract award.
- If there is a discrepancy, the more stringent or most expensive detail shall apply.
- The architect is the sole interpreter of the documents and her interpretation shall stand.

There are, of course, numerous discrepancies and resultant requests for additional funds. The contractor has submitted 50 requests for change orders and the structure is not yet complete. The developer and architect remain strong that the contractor should have assumed an answer in its estimate and should have included appropriate contingencies. All of the change orders are rejected. Halfway through the project, the contractor pulls off of the jobsite. The GC has been paid $3 million of the base contract but has not received anything for the $400,000 worth of change orders that are on the table. The GC has heard from their subcontractors that at least double this figure will be coming when they mobilize on the project. Can the contractor walk away from a project such as this? What sort of notice would be required? The owner will of course claim the contractor has improperly terminated the contract and they are being damaged for delay. Are "all-inclusive" terms such as this fair in the documents? Assuming this case goes to court, how would a judge rule?

Case 31: Seismic repairs

The owner of this multi-tenant retail facility is a 70-year-old retired construction management professor. He has invested his life savings into the building and acts as his own property manager. The facility received significant cosmetic and structural damage during an earthquake. All of the tenants were evacuated until repairs could be completed. The owner immediately employed an architect, a structural engineer, and a general contractor to implement repairs. The owner felt in this manner he was acting in the best interest of himself and his tenants. By expediting repairs, he would get his tenants back into the facility, thereby minimizing loss of sales. He felt he was also looking out for the insurance company. The design teams' corrections were well-documented and the contractor's costs were tracked on a time and material basis. The insurance company also immediately mobilized but did not take any action. They preferred to sit back and watch the corrective work take place. Upon completion of this $10 million repair project, the insurance company then questioned whether the fixes were proper and whether the actual costs reported were accurate. They also referred to a provision in the property insurance policy, which required the owner to obtain competitive bids on all portions of repair. The insurance company ultimately offered the owner 80 cents on the dollar for repairs, which the owner passed through and offered to both his design and construction teams. Will they accept? Who loses in this situation? What should the different team members have done to have assured 100% collection on the repairs? Is this a standard situation in the insurance claim arena? If the owner sues the insurance company, will he collect 100% of costs incurred by himself and his agents? Will the tenants pitch in?

_____

_____

_____

_____

_____

Case 32: Line item estimate

A software development client entered into an agreement with a general contractor to build a simple 50,000 square foot office building with an open office concept. The client's architect questioned three estimate line items in the contractor's guaranteed maximum price proposal, which appeared to be heavy. The contractor responded that these costs may be needed and the owner will receive 90% of any savings, according to the terms of the contract. The contractor did not manage the project efficiently and finished four months late. Because of this delay, the contractor reported that they overran the bottom-line GMP and there were not any savings to share. If a contractor mismanages the work, does the owner lose an opportunity for savings? The architect closely monitored the work involved with the three subject estimate items and was confident that the contractor under-ran the estimate in these areas. The contractor indicated that it was not a "line item guaranteed maximum price estimate, but rather only the bottom line was what was guaranteed." If there were savings on single line items, especially ones that were questioned early, can the contractor use these savings to offset overruns in other areas? Will an audit now help either party? What do standard contracts say?

_____

_____

_____

_____

_____

Case 33: Allowance accounting

An electronics manufacturing facility has employed a general contractor to construct a 1,000-car post-tension underground parking garage. There are three allowances included in the guaranteed maximum price contract. All three are essentially contingency funds that the contractor may need due to rain delays, poor soil conditions, and dewatering, each totaling $100,000. None of these conditions truly came to pass. When the client asked for the allowances to be change-ordered out of the contract, the contractor delayed. They argued that the contract did not state when the allowances were required to be accounted for. How and when are allowances formalized into the contract? As time went on, the contractor used creative accounting and charged all sorts of marginal costs against the allowances. Ultimately, the allowances were all "spent." How can allowances be defined and managed so that both parties are treated fairly? Should allowances exist? How is the use of allowances any different than an open-ended time and materials contract?

_____

_____

_____

_____

_____

Case 34: Contract concerns

Your firm is competitively bidding a hotel project on the waterfront. The market is slow. You have experience in this geographic area as well as the hotel construction arena. Your firm has worked with the prime architect on numerous successful projects. The architect is contracted by a national hotel chain that will operate the hotel when complete. Your contract will be with the property owner, who is retaining ownership of the dirt, which makes them a joint venture (JV) partner in the completed facility. The bid documents include a homegrown contract that appears to have been developed by the hotel's attorney. The contract references an attached AIA A201 general conditions document, which has been substantially modified. These two contract documents have several conflicts. The bid form requires your officer's signature acknowledging that you do not take any exceptions to the proposed contract agreement. You know the architect from previous projects and have had an opportunity to review their contract with the hotel operator, and discover that this is also in conflict with the construction contract documents, including the AIA A201. Do you send your contract to your attorney? What will be their advice? This is a bonded project. Do you send the contract to your bonding agency? What will be their advice? Do you pull out of the competition? Do you submit a qualification with your bid stating all of your contract concerns and recommended changes? What will your competition do? Do you stay silent to all of this and choose to fight it out after the award, or wait until an issue arises during the course of the project?

_____

_____

_____

_____

_____

4

ESTIMATING

Introduction

Cost is one of the most critical project attributes that must be controlled by the project manager and superintendent. Construction costs are estimated to develop a budget within which the jobsite team must build the project. All project costs are estimated in preparing bids for lump sum or unit price contracts and negotiating the guaranteed maximum price on cost-plus contracts.

Cost estimating is the process of collecting, analyzing, and summarizing data in order to prepare an educated projection of the anticipated cost of a project. Project cost estimates may be prepared either by the project manager or by the estimating department of the construction firm. Even if the estimate is prepared by the estimating department, the project manager must understand how the estimate was prepared because he or she along with the superintendent must build the project within the estimate. This project budget becomes the basis for the cost control system, as will be discussed in the Chapter 10 case studies.

There is no one exact estimate for any project. There are many correct estimates, although some are more correct than others. Adjustments in pricing, subcontractor and direct labor strategies, overhead structures, and fee calculations are individual contractor decisions that will determine the "best estimate" for those conditions at that time. The process of developing an estimate is similar to constructing floors in a skyscraper, as illustrated in Table 4.1. The first floor of gathering information is the foundation for the process. As the estimator proceeds through the process, information continues to be analyzed and summarized and refined, until eventually there is only one figure left, the final estimate or bid.

TABLE 4.1 Estimating process

Final bid or proposal
Receive bid day subcontractor and supplier quotes
Determine markups
Prepare bid form
Refine ROM
General conditions estimate
Summary schedule
Subcontractor plug estimates
Direct work pricing recaps
Quantity take-offs for direct and subcontracted work
Initial calls to subcontractors
Assign estimating team responsibilities
Work breakdown structure
Develop an early ROM
Visit the site
Attend pre-bid meeting
Initial project overview

Types of cost estimates

Some use the term "estimate" to be all-encompassing. There are several different types of cost estimates, and confusion between these may cause conflicts between the parties, as represented by case studies throughout this book. *Conceptual cost estimates* are developed using incomplete project documentation. A type of conceptual cost estimate called a rough order of magnitude (ROM) estimate is developed from the first document overview to establish a preliminary project budget and determine if the contractor intends to pursue the project. *Detailed cost estimates* are prepared using complete drawings and specifications, and are usually associated with lump sum, stipulated sum, or bid projects. *Semi-detailed cost estimates* are used for guaranteed maximum price (GMP) contracts, and have elements of both conceptual and detailed estimates. All estimates have the following major elements or categories, some of which require a more detailed effort by the estimator than others:

- direct labor;
- direct material;
- construction equipment;
- subcontractors;
- jobsite general conditions; and
- markups, including fee.

Similar to constructing a building, estimating is a series of steps, the first being project overview to determine if the project is going to be pursued. The quantity take-off step is a compilation of counting items and measuring volumes. Pricing is divided between material pricing, labor pricing, and subcontract pricing. Labor is computed using productivity rates and labor wage rates. Material and subcontract prices are developed most accurately using current local supplier and subcontractor quotations. Even though most contractors consider direct labor as their riskiest portion of the estimate, as shown in several of the case studies in this chapter, proper analysis of subcontractor bids is also very important to first landing a project and then successful performance. Jobsite general conditions cost is a job cost and is largely schedule-dependent.

Home office indirect costs are often combined with desired profit to produce the fee. The fee calculation varies dependent upon several conditions including company volume, market conditions, labor risk, and resource allocations. Other markups include labor burden, liability insurance, excise tax, liability insurance, and potentially performance and payment bonds. An example estimate summary is included as Figure 4.1.

King County Builders, Inc.
4100 SW Hilltop Road
Rainier, WA 98111
206-527-4343

GMP SUMMARY ESTIMATE
Health and Wellness MOB, Project #869

CSI Div	Description	Direct L. Hours	Labor	Material	Equipment	Subs	Total
01	Jobsite GCs @7.5%	0	$344,211	$250,750	$325,000	$0	$919,961
02	Demolition	NIC	NIC	NIC	NIC	NIC	NIC
03	Concrete	3,100	$127,100	$204,400	$15,000	$0	$346,500
04	Masonry	0	$0	$0	$0	$110,000	$110,000
05	Structural & Misc. Steel	10,618	$414,120	$525,000	$147,500	$0	$1,086,620
06	Rough and Finish Carpentry	2,114	$88,800	$128,500	$0	$0	$217,300
07	Thermal and Moisture	0	$0	$0	$0	$650,000	$650,000
08	Doors and Windows	2,929	$123,000	$121,000	$0	$850,000	$1,094,000
09	Finishes		$0	$0	$0	$1,652,010	$1,652,010
10	Specialties	238	$10,000	$25,000	$0	$0	$35,000
11	Equipment	1,875	$75,000	$375,000	$0	$120,000	$570,000
12	Furnishings	NIC	NIC	NIC	NIC	NIC	NIC
13	Special Construction	0	$0	$0	$0	$0	$0
14	Elevators	0	$0	$0	$0	$355,000	$355,000
21	Fire Protection	0	$0	$0	$0	$215,000	$215,000
22	Plumbing	0	$0	$0	$0	$1,030,000	$1,030,000
23	HVAC and Controls	0	$0	$0	$0	$1,344,500	$1,344,500
26	Electrical	0	$0	$0	$0	$1,520,220	$1,520,220
27	Low-Voltage Electrical	0	$0	$0	$0	$420,000	$420,000
31	Sitework	1,389	$50,000	$75,000	$25,000	$1,470,000	$1,620,000
	Subtotals Direct Work:	22,263	$888,020	$1,453,900	$187,500	$9,736,730	$12,266,150
	Totals DW&GCs:	22,263	$1,232,231	$1,704,650	$512,500	$9,736,730	$13,186,111
	Labor Burden on Direct L	50.00%	$888,020	$444,010			$13,630,121
	Labor Burden on Indirect L	30.00%	$344,211	$103,263			$13,733,385
	Subcontractor Bonds	w/Sub $		$0			$13,733,385
	State Excise, B&O Tax	0.75%		$103,000			$13,836,385
	Liability Insurance	0.95%		$131,446			$13,967,831
	Builder's Risk Insurance	by Owner		$0			$13,967,831
	Fee	5.00%		$698,392			$14,666,222
	Contingency	1.50%		$219,993			$14,886,216
	GC Bond	Excluded		$0			$14,886,216
	Total GMP:						**$14,886,216**

FIGURE 4.1 Summary estimate

There are many good textbooks dedicated solely to the topic of estimating, including *Construction Cost Estimating* by Holm and Schaufelberger. This chapter is only a brief introduction. The interested reader should look to one of those books for a more detailed coverage. The three major lessons to be learned in estimating is first to be organized. If proper organization and procedures are utilized, good estimates will result. The second is to estimate, and estimate a lot. Practice and good organization will eventually develop thorough and reasonable estimates. The third is to maintain and use historical databases of cost and productivity rates from previous projects.

Case studies

Case studies in this chapter include:

35. No subcontractor coverage
36. Subcontractor short list
37. Unknown electrician
38. Estimating too smart
39. Window bids

Most of these case studies overlap with other primary topics, and almost 25% of the cases in this book involve some aspect of estimates. Some other cases that are directly connected include 7, 9, 12, 24–33, 46, 51, 57, and 89.

Toolbox quotes

The only way to become an estimator is to estimate, and estimate a lot!

$$\begin{array}{r} \textit{Direct labor} \\ + \textit{ direct material} \\ + \textit{ subcontractors} \\ + \textit{ indirect costs} \\ + \textit{ equipment} \\ + \textit{ markups} \\ \hline = \textit{ your bid} \end{array}$$

There is no such thing as the "correct" estimate.

Case 35: No subcontractor coverage

A Portland, Oregon, pharmaceutical firm is letting a project to bid to a select list of four qualified general contractors. The project will ultimately be awarded for $27 million, which is approximately $1,000 per square foot. This is very complicated work. The architect is very experienced. The client's local West Coast office is utilizing the services of its corporate owner's representative from the East. This individual has a reputation of beating up the entire construction industry, including general contractors (GCs), subcontractors, designers, and even the city. This will be the fourth project for this client locally, and none of the local GCs have worked for them on a second project. Contractor 1 does not turn in a bid. Contractor 2 is involved in litigation with the owner from a previous project and the owner essentially disqualifies their bid. Contractor 3 violates the bid rules and is disqualified. Contractor 4 did not really want the project due to their busy workload and the client's combative reputation, but turned in a courtesy bid. Their estimator was a "short-timer" and resigned the day after the bid was posted. Contractor 4 was surprised to receive a phone call from the client notifying them of intent to award. The estimator responsible did not pursue the subcontractor industry for quotations. Because of the client's reputation, many of the qualified subcontractors did not pursue the project. Thirty different specification sections received one or zero quotations and were plugged by the estimator.

a. What errors did contractor 4 make in the estimating process? Should they accept the award? Can they ethically go out and rebid to the subcontractor industry? Are they in a strong or a weak position for subcontractor and supplier buyout opportunities? Who has the upper hand with contract negotiations in this scenario, the contractor or the owner?

_____

_____

_____

_____

_____

b. Because the GC did not anticipate getting the project, and the estimator distanced himself from it as soon as possible, they had to go outside to hire a contract project manager (PM). This PM had a similar reputation to the owner's representative. He was known to be extremely tough, and the GC felt this is the type of personality they needed to financially come out even, if not ahead. How do you think the two individuals approached doing business with each other? How did they deal with the other team members? Would you be surprised to find out that this project was a huge success?

_____

_____

_____

_____

_____

Case 36: Subcontractor short list

An outside third-party project manager (also known as an owner's representative) working for a public university solicits bids from three general contractors with the stipulation that the major subcontractors must be chosen from the client's pre-approved list. The successful general contractor (GC1) awarded the electrical scope to a firm that was not on the list. The client's representative is aware of this situation but remains quiet. The second-place general contractor (GC2) finds out about this infraction and files a complaint. Should the university stay with the original award? Should the project be rebid by GC1 to the select electrical firms? Should the project be awarded to GC2? Should the entire bid process be thrown out and started over? Is this fair to the taxpayers? Is it fair to the low electrical firm? Can either the low general contractor or the low electrical subcontractor file a complaint if they are not allowed to proceed with the project? Does this situation make a good case to disallow short-listing on public works projects? What should the owner's representative have done to prevent this from happening? Should he be terminated?

_____

_____

_____

_____

_____

Case 37: Unknown electrician

For the last four weeks, three individuals in your firm plus yourself have worked full-time estimating a $50 million public bid hospital project. Your volume is very low and you need this project. You figure that your office is easily $25,000 in the hole already with estimating expenses. On bid day, you receive an unsolicited bid from an out-of-state electrical subcontractor. Their price is $750,000 below the lowest local firm. You have never worked with this firm before and do not know anything about their qualifications. The local electrical inspector has a reputation of being very tough. With only 10 minutes to go before the bid is due, your office calls the low electrical subcontractor, but their phones are jammed and you cannot get through. Their emailed bid was generically addressed to your firm and simultaneously to your four major competitors. What will your competition do? What will you do?

_____

_____

_____

_____

_____

Case 38: Estimating too smart

You are a contracted owner's representative for a new electrical trade's educational facility. At the request of the board of trustees, you have put this relatively simple $2 million project to bid to a short list of four general contractors. You are very familiar with three of the GCs, who are of comparable size and experience. The fourth is an out-of-town firm who is substantially smaller, but has constructed these types of facilities in the past. Two of the larger general contractors are relatively silent throughout the bid cycle. They diligently pick up their drawings, attend the pre-bid meeting, and appear to be earnestly estimating the project. The third larger firm asks over 100 detailed questions about the documents. The estimator is a friend and you advise him that he is becoming "too smart" on the job. All of his questions are answered in writing and issued as addenda to all four general contractors. The smaller contractor does not attend the pre-bid meeting. His estimator calls you with only 10 days left before the bid is due and wonders where he can pick up drawings and when the bid will be due. He does not ask any detailed questions. The morning the bid is due, he asks for a time extension but is denied. As anticipated, the detailed estimator turns in the high bid. The two silent contractors are in the middle. The smaller general turns in a bid that is $100,000 lower than the others but is within 5% of the total. The client is very delighted about the tight bid results and with the low bidder's qualifications. What should your advice be? Is it ethical to disqualify someone who was on a short list? If hired, what will eventually happen with this low firm? Is this fair to the contractors who prepared what are anticipated to be "complete" bids? Which firm will prepare the fewest change orders?

Case 39: Window bids

You have received three window wall subcontract bids. The low bid of $500,000 was from an unsolicited subcontractor. They did not fill out your prescribed bid forms. They neither acknowledge nor deny the bid documents or addenda. You know nothing about this firm. Their bid does not state any exceptions or qualifications regarding their proposal. The second bid of $600,000 is from a glass firm that you have worked with prior on other sites, but not successfully. The quality of their work was fine, but they battled with your office with respect to contract issues. They were short-listed and requested to bid on this project. They filled out your bid form, but their proposal came with an extensive list of exclusions, assumptions, and qualifications. The third bid was also from a local glazing firm whom you had solicited a proposal from and have worked with successfully. Their bid is exactly per your prescribed bid form. They do not state any qualifications or exceptions to the bid documents. Their bid is $700,000. Your pre-bid budget was $600,000 for this area of work. What do you do? How would your answers differ if this were: (a) a lump sum competitively bid job; or (b) a negotiated GMP job?

5

SCHEDULES AND SCHEDULE CONTROL

Introduction

The contract schedule is a document and a project management tool just as is the estimate. Time management is just as important to project success as is cost management. The key to effective time management is to carefully plan the work to be performed, develop a realistic construction schedule, and then manage the performance of the work. People will often use the terms "planning" and "scheduling" together, but they are different processes. Planning is the upfront work that makes the schedule feasible. Planning is a process and the schedule is the result. The schedule is a logical arrangement of activities in order of occurrence and prerequisites, and is often charted with a timeline.

Schedules are important tools for all members of the owner, design, and construction teams. Proper planning of the project and the schedule, with input from the relevant personnel such as the superintendent and the subcontractors, are keys to developing a useful construction management tool. Schedule development begins with proper planning that considers many variables such as deliveries, logic, manpower, and equipment availabilities. There are many different types of schedules, each of which has a use on a construction project. Some of the major ones include:

- detailed schedule – may be the contract schedule;
- summary schedule – may be the contract schedule;
- three-week look-ahead schedules;
- expediting and submittal schedules;
- specialty schedules – include those focused on one area of the building or phase, or on just one subcontractor; and
- pull planning schedules – part of lean construction, as will be discussed in Chapter 10.

The schedule has important contractual implications, and should be referenced and incorporated into the prime contract, and each subcontract as a contract exhibit. Liquidated damages (LDs) may be imposed if a contractor finishes the project late, as discussed in case studies in this chapter. If LDs are required, they must be clearly spelled out in all contractual agreements.

Schedules are working documents that need updating as conditions change on the project. For example, contract change orders generally modify some aspect of the scope of work requiring an adjustment to some part of the schedule. In addition to aiding management of the project, updated schedules can also be used to justify additional contract time on change orders and claims. Updating or revising the schedule can be an expensive task, and does not need to be a monthly occurrence if the project remains on track. Using the schedule to monitor progress is the superintendent's responsibility, and occurs during the weekly owner-architect-contractor (OAC) coordination meeting.

Schedule control involves monitoring the progress of each scheduled activity and selecting appropriate mitigation measures to overcome the effects of any schedule delays. As-built schedules are prepared to allow project managers to develop historical productivity factors for use on future projects. Controlling the schedule of subcontractor progress can at times be problematic for general contractors, as shown in some of the following case study examples. An example summary schedule that could be attached as a contract exhibit or included with a negotiated contractor's proposal is included as Figure 5.1. Planning and scheduling, similar to

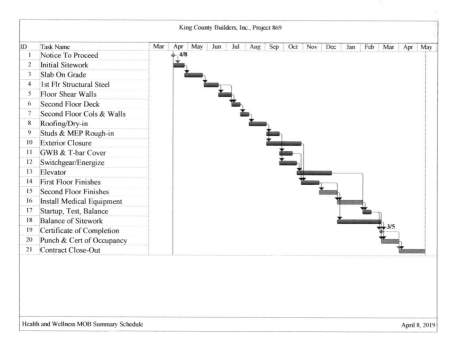

FIGURE 5.1 Summary schedule

estimating, have been covered extensively in other books dedicated solely to that topic. Most project management books such as *Management of Construction Projects* and *Introduction to Construction Project Engineering* also devote at least one chapter to scheduling. This chapter is just a brief introduction, and the example cases are indicative of some of the scheduling hurdles the jobsite team must overcome.

Case studies

Case studies in this chapter include:

40. Glazing schedule
41. Drywall subcontractor
42. Liquidated damages
43. Schedule hold

Most of these case studies overlap with other primary topics. Case studies 17, 22, 49, 56, 91, 92, and 93 also involve schedules.

Toolbox quotes

Status or update, but avoid revising the schedule.

Schedules, like estimates, should be contract documents.

A schedule is an important project management tool; it should take on whatever format necessary to make it an effective tool.

Case 40: Glazing schedule

You, as the general contractor's (GC's) project manager (PM), have a problematic glazing subcontractor. They are behind schedule. They refuse to work overtime to catch up. The subcontractor has submitted several unsubstantiated change order proposals (COPs) that have not yet been approved and they are threatening to stop work. They have switched out both the project manager and the superintendent since the project started. They are not staffing the project according to their planned and committed manpower. You are not getting along personally with the subcontractor's current project manager and have resorted to communicating only through email. You are receiving pressure from the field to resolve the problem and get the glazier to perform. Your supervisor has indicated that it is your responsibility to solve the problem. What do you do? What could you have done to prevent these problems from occurring? What recommendations can you make to a general contractor's subcontractor management system to prevent these types of situations?

———————————————————————————————————
———————————————————————————————————
———————————————————————————————————
———————————————————————————————————
———————————————————————————————————

Case 41: Drywall subcontractor

Your drywall subcontractor is not performing in the field. They have not staffed the project according to their original commitment or to your field superintendent's expectations. They are holding up the work of other trades. The quality of the subcontractor's completed work has been unacceptable, and you are constantly on them to improve. You have asked for the removal of their superintendent, but the firm has refused. Can you contractually require a subcontractor to change personnel? Can they contractually refuse? It is eventually decided by your home office that the subcontractor must be terminated. How do you go about this process? Is it simple? How is it documented? Will your firm get sued for false termination? What does standard contract language say? How do you protect yourself? Will it be easier to just keep limping along with them? Would your answers differ if your GC firm had either a lump sum or a negotiated contract with your client?

———————————————————————————————————
———————————————————————————————————
———————————————————————————————————
———————————————————————————————————
———————————————————————————————————

Case 42: Liquidated damages

Your client has assigned $2,000 per day liquidated damages for late completion to your $25 million turnkey contract. You in turn have passed these liabilities onto your subcontractors. You have one exterior siding subcontractor who sends you written notices requesting additional time whenever a request for information (RFI) or a submittal is a day late being returned. They document every adverse weather day. They document when other subcontractors are holding them up. They request schedule extensions with every change order. Not all of their documentation is substantiated, but some of it may be. They are claiming a total of 20 additional workdays. This is an administrative nightmare for you.

a. How do you deal with the siding contractor during the course of the project? Do you pass through these notifications to your client? If this subcontractor ultimately finishes behind schedule by 10 days, and you also finish behind schedule, does your client collect from you? Do you collect from the subcontractor? How does it get resolved typically? What is the contractual and legal resolution? Analyze both ways: (a) with stipulated LDs in the subcontract agreement; and (b) without.

_____

_____

_____

_____

_____

b. Do you offer a subcontractor a bonus (dollars per day) for finishing early? Isn't this only fair? If the penalty were $2,000 per day for finishing late, what would the bonus be? Assume the siding contractor above finished per their original schedule, but had built up the claimed additional 20 days, and had such a bonus clause. They would now claim that a bonus was due for the $40,000. Do they get it? If not, why not? If so, who pays, you or your client? How is this resolved?

_____

_____

_____

_____

_____

Case 43: Schedule hold

a. Your construction company does $200 million in annual volume with a home office overhead cost amounting to a total of 2% of annual revenue. This particular 16-month project has costs estimated at $20 million plus 8% jobsite general conditions costs plus a 4% fee. Ignore other markups for this exercise. Assume that at exactly the midpoint of the schedule, your project was put on hold for one month due to reasons beyond the contractor's control such as weather, union strikes, owner financing, or city issues. Pick one. Using the contract, classes, this book, and research outside of the classroom, how should the contractor properly deal with this delay? Discuss issues such as notice, documentation, jobsite administration costs, home office costs, loss of fee, loss of productivity, and quality and safety concerns. What is the "Eichleay Formula"? Prepare a claim for this delay.

_____

_____

_____

_____

_____

b. As an alternative to the claim preparation above, prepare a recovery plan to get the owner their building on time. Show with the schedule that a recovery is possible. Submit a change order proposal for the anticipated costs associated with working in an expedited fashion. Provide all necessary cost backup.

_____

_____

_____

_____

_____

c. Assume this time that the one-month delay is your fault, the general contractor, through the actions of your site utility subcontractor. The underground fire loop was installed and backfilled without requesting inspection from the city. The city subsequently required the entire system to be re-exposed, which seriously impacted all of the work on the project due to lack of access. From the general contractor's perspective, how do you deal with the subcontractor? How does this impact other subcontractors? How do you respond to the impact this delay has caused the owner?

_____

_____

_____

_____

_____

d. From the owner's perspective how do you deal with the real costs that the contractor-caused delay has placed upon you? Do you have any recourse contractually? Are consequential damages recoverable contractually? Compare the differences and philosophies (advantages and disadvantages for all parties) between LDs and actual or real damages.

_____

_____

_____

_____

_____

6

SUBCONTRACTORS AND SUPPLIERS

Introduction

Commercial general contractors use subcontractors to execute most of the construction tasks involved in a construction project. A typical general contractor (GC) today subcontracts 80–90% of the work. Subcontractors, also referred to as specialty contractors, therefore are important members of the general contractor's project delivery team and have a significant impact on the GC's success or failure. Since subcontractors have such a great impact on the overall quality, cost, schedule, and safety success for a project, they must be selected carefully and managed efficiently. There must be mutual trust and respect between the superintendent and the subcontractors because each can achieve success only by working cooperatively with the other. Consequently, project managers and superintendents find it advantageous to develop and nurture positive and enduring relationships with reliable subcontractors. Project managers and superintendents must treat subcontractors fairly to ensure they remain financially solvent not only to finish this project, but to be available to provide competitive bids on future projects.

One of the reasons general contractors use subcontractors is to reduce some of the risks inherent in construction. One of the major risks in contracting is accurately forecasting the amount and cost of direct craft labor required to complete a project. By subcontracting significant segments of work, the general contractor can transfer much of the risk to subcontractors. When the project manager (PM) asks a subcontractor for a price to perform that scope of work, the subcontractor bears the risk of properly estimating the labor, material, and equipment costs. Subcontractors have access to specialized skilled craftsmen and equipment that may be unavailable to the GC. Craftspeople experienced in the many specialized trades required for major construction projects are expensive to hire and are generally used on a project site only for limited periods of time. It would be cost-prohibitive for a GC to employ all types of specialized labor full-time as a part of its own workforce.

Subcontracting reduces, but does not eliminate, all risks for a general contractor. The project manager and superintendent give up some control when working with subcontractors. The scope and terms of the subcontract define the responsibilities of each subcontractor. If some aspect of the work is inadvertently omitted, the GC is still responsible for ensuring the contract requirements are achieved. Specialty contractors are required to perform only those tasks that are specifically stated in the subcontract documents. Consistent quality control may be more difficult with subcontractors, as shown in a few of these case studies. Project owners expect to receive a quality project and hold the GC's superintendent and project manager accountable for the quality of all work, whether performed by the general contractor's direct crews or by subcontractors. Subcontractor bankruptcy is another risky aspect of subcontracting, which can be minimized by good prequalification procedures and timely payment for subcontract work. Scheduling subcontractor work often is more difficult than scheduling the general contractor's crews, as was shown in the previous chapter. Safety procedures and practices among subcontractors may not be as effective as those used by the general contractor, presenting an additional challenge to the superintendent. All of these financial and control aspects are even more crucial when subcontractors subcontract portions of their work to third-tier subcontractors and suppliers, as shown in Figure 6.1.

Subcontractors and suppliers are discussed in many of the case studies throughout this book. The case studies in this chapter will discuss subcontractor selection, acquisition, and management. Selecting quality "best-value" subcontractors is essential if the project manager and superintendent are to produce a quality project on time and within budget. Project managers must remember that poor subcontractor performance will reflect negatively on their professional reputations and their ability to secure future projects. Once the subcontractors have been selected, contract documents are executed documenting the scopes of work and the terms and conditions of the agreement. Subcontractor management is an integral part

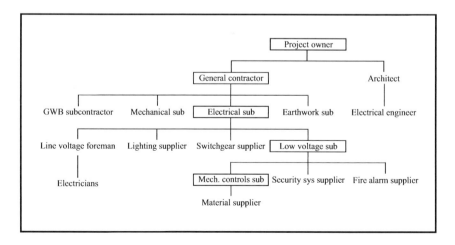

FIGURE 6.1   Organization chart

of project management. While the project superintendent manages the field performance of the subcontractors, the project manager manages all subcontract documentation and communication, and is responsible for ensuring the subcontractors are treated fairly.

Case studies

Case studies in this chapter include:

44. HVAC union
45. Union general contractors
46. Hospital buyout
47. Young engineer
48. Carpet bankruptcy
49. Steel supplier
50. First team
51. Concrete walls
52. Pulled quote
53. Hostile project manager
54. Fire protection heads

Most of these case studies overlap with other primary topics, and almost one-third of all of the cases in this book involve subcontractors and suppliers to some extent. Additional case studies that are directly related to subcontractors and suppliers include 12, 15, 29, 35–37, 38–42, 56, 61, 65–68, 76, 79, 85, 86, 89, 91, 92, and 93.

Toolbox quote

Subcontractors: you can't live with them, and you can't live without them.

Case 44: HVAC union

This general contractor has chosen a union heating, ventilation, and air conditioning (HVAC) subcontractor on a predominantly open shop project. The HVAC subcontractor is very qualified, was the low bidder, and so far is performing well. But unfortunately, some of the crew begin intentionally damaging roof flashing and louvers, which is under contract of the open shop roofing subcontractor. Then, they intentionally slow down their own work, which is slowing down work of other trades. Eventually, the "tinners" (nickname for sheet metal workers) pull off the job. The union sheet metal trade is picketing the job and bothering other craftsmen who cross the picket lines. These actions are all from the crew, not the

HVAC subcontractor. This is costing the subcontractor money as well. How did this happen? Who in the general contractor's organization is responsible to solve the problem? Can you as the general contractor's project manager contractually force the subcontractor to replace the crew and/or return to work? How can you prevent this from happening in the future?

_____

_____

_____

_____

_____

Case 45: Union general contractors

Many general contractors are signatory to the carpenter and laborer unions. A few others also have agreements with trades such as cement finishers, ironworkers, operating engineers, and teamsters. If a contractor is signatory to one but not all unions, are they considered to be a "union contractor"? Are union GCs required to employ subcontractors such as electrical or plumbing who have other union affiliations? Do union subcontractors feel that a union GC should only employ all-union subcontractors? What is customary? Do union subcontractors only work for union GCs? Why the difference? Does a carpenter-signatory general contractor only employ union drywall subcontractors who also employ carpenters? What would be the rule if the union general contractor subcontracts work out that they normally self-perform, such as horizontal formwork for elevated concrete slabs performed by carpenters? Can they do this open shop? Do market conditions or geographic locations affect any of this?

_____

_____

_____

_____

_____

Case 46: Hospital buyout

During buyout of a $40 million hospital expansion bid project, you, as the GC's PM, are faced with the following possibilities:

a. You can package three specification sections (drywall, insulation, and paint) all under one $2 million subcontract. This will all be work that the subcontractor will self-perform. Before you make this decision, you should consider the following three scenarios: (1) If the packaging results in a slight cost increase of $10,000, does this reduce your work effort, and is it worth it? (2) Conversely, if the packaging results in an overall cost savings of $50,000 to the general contractor, how can you be assured that you are not overloading the subcontractor? (3) Finally, if it is a push in cost, is it a good idea or a bad idea, and why?

_____

_____

_____

_____

_____

b. You received a bid from a qualified concrete reinforcement steel (rebar) installation subcontractor for a portion of the work, which your firm normally self-performs with your own ironworkers. It is $100,000 less than you have estimated. (1) List three reasons why you should hire this subcontractor. (2) List three reasons why you should not hire this subcontractor.

_____

_____

_____

_____

_____

c. Assuming that you did hire a subcontractor for work that you normally self-perform, you are now faced with stiff criticism from your field supervision because you have taken work away from his "guys and gals." The subcontractor is not performing to your superintendent's satisfaction. He is on your back continually, requesting you to terminate the subcontractor. Why might you have thought this was a good idea? How do you now deal with this issue?

_____

_____

_____

_____

_____

d. Your client has requested that you solicit a bid from, and do your best to employ, his favorite painter. Your firm has never worked with them prior. Should you include the subcontractor on the list? If the subcontractor is second bidder by $25,000, and the owner agrees to pay the difference, should you hire them? Is this ethical? If you proceed and the painter does not perform, what responsibility, if any, does the client share?

_____

_____

_____

_____

_____

e. You have a bid from a floor covering subcontractor who is a "broker"; they do not fabricate any materials or perform any direct field labor. The contractor will buy the material from an outside supplier (or two or three), and will hire separate individual subcontractors for installation of carpet, vinyl, and ceramic tile. The subcontractor's price of $400,000 is $20,000 lower than the second bidder. Is this to your advantage? Why or why not? What is your company's position to this arrangement? How do you keep the prime subcontractor from stepping back and allowing you to manage their third-tier subcontractors and suppliers directly?

_____

_____

_____

_____

_____

f. Your senior project manager has directed you to employ a subcontractor who will buy the doors, door hardware, and door frames, something you normally purchase directly, and have them installed, all in a $175,000 package deal. This will be $7,500 more than if you purchase the materials from three separate suppliers and hire an installer. Is this to your advantage? Why or why not? At what premium is it not to your advantage? Draw an organization chart to reflect this scenario. Show contractual connections as well as lines of communication.

_____

_____

_____

_____

_____

g. If your firm decided not to package the door system described above, what can you do to ensure that all of the work will be performed properly by these four companies? If the doors and frames are not delivered machined properly, who is at fault? Who checks the shop drawings and the submittals? Who pays? Is there a back charge procedure to follow? How do you solve the problem in the field? Draw an organization chart to reflect this scenario.

_____

_____

_____

_____

_____

h. You have an opportunity to break apart a $4 million mechanical bid proposal that is normally purchased as a package. The prime HVAC sub has offered you this opportunity to save its markup of $200,000 on managing these multilayered subcontractors. This includes separate fire protection, controls, mechanical insulation, plumbing, and pipe labeling subcontractors. Take the position that you will accept this deal. Your senior project manager recommended against it but allowed you the rope to succeed. Why did you feel this was to the general contractor's advantage? How will you manage the work so that you personally succeed? If it becomes a problem during the project, how can you correct it? Draw an organization chart for this scenario.

---

i.   Take the opposite position to the mechanical buyout described above. Draw an organization chart packaging all of the mechanical trades under the HVAC subcontractor. Is this the correct arrangement? Do they manage their sub-trades adequately or will you have to? Are you, as the general contractor's project manager, allowed to communicate directly with the HVAC subcontractor's subcontractors? Will joint check agreements be required? At what savings is it worth splitting this work apart?

Case 47: Young engineer

You are responsible for two subcontractors, earthwork and roofing, on your megaproject. You are a 23-year-old field engineer (FE). Experienced 40- and 50-year-old project managers work for both of these subcontracting firms. Both of these subcontractors are very qualified and have worked with your firm on previous occasions. The firms are financially strong and their field work is performed safely and is of acceptable quality. Your problem: the Rodney Dangerfield Syndrome, "I don't get no respect!" These subcontractors continually go over your head to your project manager. They talk to your peers on other projects. They do not respond to your requests for information or requests for additional backup on change orders. What do you do now? Do you care? How do you establish yourself as the point of contact for these firms? List five rules of order you would recommend for a beginning engineer who needs to earn the respect of experienced subcontractor managers who work for him or her.

Case 48: Carpet bankruptcy

Your firm has not worked with this $500,000 carpet subcontracting firm before. You, as the GC's PM, originally decided to bond them, but were surprised that their bond rate was 4% of contract value; most subcontractors are between 1% and 2%. They have submitted several change order proposals for increases in scope as well as discrepant design documentation. Most of these are pending processing due to one reason or another. The quality of their work is fine. They are supporting your schedule. You have recently received multiple material men's notices and pre-lien notices from the carpet subcontractor's third-tier suppliers. The subcontractor has indicated that they are solvent and do not have any cash flow problems. You have just discovered that the chief executive officer of their firm has left. You have made a few calls and have caught wind that they may be going bankrupt. What do you do? What contractual recourse do you have? Should you hold up future payments? Should you notify your client? Will the carpet subcontractor's third-tier suppliers come after you for payment? Can they? Assuming the subcontractor sells their receivables to a collection agency, will that firm also come after you? What adjustments to the system can you recommend to prevent this situation from happening in the future?

Case 49: Steel supplier

Your company has never worked with this structural steel fabricator prior. They have been marketing to your home office staff purchasing agent, and he has decided to give them a try and issued them a purchase order and handed it off to you, as the GC's PM, to manage. This is a very complicated steel detailing project. The fabricator has subcontracted the detailing out to a third-tier firm. There are numerous questions and meetings with the detailer. The shop drawings show up relatively on schedule but require a substantial amount of your time for review. There are numerous changes that are brought up during the shop drawing process. Many of these will ultimately result in justifiable change orders to the client. In short, you are totally engulfed in the detailing effort. The fabricator has invoiced you for detailing, as well as the complete steel mill order material for purchase and some fabrication. You have been paid and have paid them accordingly. You had no reason to suspect that anything was wrong. The embedded steel members show up on time and have been fabricated properly. You have not visited the fabricator's shop. Your first major shipment of columns and beams is scheduled for next Thursday. On Monday, your senior project manager discovers that you had not inspected the steel during fabrication, and suggests you take a trip to the supplier's shop. On Tuesday, you show up at the shop unannounced and cannot find any of your steel. The fabricator tells you it is temporarily being stored at the rolling mill. You visit the mill and find out that not only is the fabricated steel not there, but the mill has not received a purchase order for your project. In fact, the mill will not do business with your fabricator due to lack of payment on a previous project. But the mill has the shapes and lengths you are looking for in stock. On Wednesday, you receive a bankruptcy notice from your fabricator.

a. How did this happen? What should you have done to prevent it from happening? Who in your organization is at fault?

_____

_____

_____

_____

_____

b. Do you try to keep the fabricator afloat? Do you buy the steel from the rolling mill directly? Do you put your people in their shop to fabricate the steel? Do you employ their personnel and pay them directly to perform the fabrication?

_____

_____

_____

_____

_____

c. You decide to hold any additional payments that are already in the system from the steel fabricator. There is some money due them for detailing and embeds. The supplier and detailer both file liens on your project. Are they valid? How do you get them removed? Your client discovers the situation and begins to pressure you to have the liens removed. Can the client contractually withhold all further payments from the GC until this is resolved? How could you have prevented this from happening?

_____

_____

_____

_____

_____

d. How do you get the steel delivered to the project and still keep your project on schedule to avoid the $1,000 liquidated damages (LDs) that are in your contract? Do you negotiate with the second bidding supplier? Do you rebid the steel package to other fabricators? Assume that the market has gotten tighter and all of the fabricators have plenty of work. Do you hire the second bidder lump sum, time and material (T&M), or unit price?

_____

_____

_____

_____

_____

e. Assume you find another fabricator. Will they want to re-detail the project? Will they accept the detailing that already exists? Which is the best situation for you? What are your risks either way? How do you mitigate these risks?

_____

_____

_____

_____

_____

f. Your other subcontractors catch wind of this problem and begin sending you notices of potential impact costs associated with delays. How do you deal with these subcontractors? Are you up front or do you try to hide the problem?

_____

_____

_____

_____

_____

Case 50: First team

A client has selected a general contractor to perform preconstruction services and to eventually enter into a negotiated contract to build a $20 million shelled office building. The construction market is very busy. Many good subcontractors are turning away opportunities to do work. The people who are now managing (office) and supervising (field) construction projects in the subcontract arena are not typically the first-team selections. During a busy market, many project engineers (PEs) are promoted to project managers, and foremen likewise to superintendents. Predict how this project could turn out if the general contractor posts the project on public plan center websites to receive open market bidding from subcontractors and suppliers. As a general contractor, it is your duty to protect the client from these risks. Prepare a list of several project management tools that could be used to mitigate both the general contractor's and the client's risks.

_____

_____

_____

_____

_____

Case 51: Concrete walls

Your company was selected as the GC on a 10-story office building project. Your contract includes many generic terms such as "contractors," "work," and "documents." It does not differentiate between a GC and its subcontractors or specific items of work. You bid out the concrete package and have three tight subcontractor bids within 3% of each other. The scope includes concrete formwork, rebar installation, concrete placement, and slab finishing. Your firm will carry purchase of concrete and concrete pump rental. You also estimated the concrete work yourself and were within 3% of the low bidder, just pennies different from the high-bidding subcontractor. Before awarding to the apparent low bidder, you decide to interview them to make sure they have the scope and schedule covered. Their written bid did not reference any drawings. In the interview, you ask if they have all concrete shown on both the architectural and structural drawings. You ask this question because there were a couple of concrete housekeeping pads and one set of steel stair infills shown on the architectural drawings that you want to make sure were picked up. They answer yes to the question. This was documented in the pre-award meeting notes. These notes were not made an exhibit to the subcontract agreement.

You decide to hire the interviewed subcontractor for the concrete work. One week later, you mail them a subcontract agreement and at the same time they start the work. Two weeks later, you get a call from the subcontractor inquiring why the reference was made to the architectural drawings. You remind the subcontractor's project manager of the pre-award meeting and the concern that he picks up the housekeeping pads and stair infills. He continues working. Two weeks later, they sign and mail back the subcontract but cross out (properly initialed) the reference to "concrete work shown on architectural drawings." As it turns out, there are several concrete walls that are shown on the architectural drawings but are not shown on the structural drawings. The walls are not detailed as they would have been on the structural drawings, which would typically include steel embeds and rebar. A request for information (RFI) is written and the structural engineer responds with the necessary sketches, but the architect does not attach a directional document such as a construction change directive (CCD) to the RFI response and the sketches. You forward this information to the subcontractor, directing them to proceed, and again remind them of the preconstruction discussion you had, including another copy of the meeting notes.

a.   The concrete subcontractor refuses to accept the subcontract as originally written, and will not install these walls unless a change order is issued. The walls were not specifically discussed in this prior meeting. They continue working. The walls are worth about $50,000. You do not have the money in your budget to take care of the issue and neither does the subcontractor. What can you do? What went wrong, and how could it have been prevented? If this were a lump sum project, would this be a change order to the owner? Would it be accepted?

_____
_____
_____
_____
_____

b. Assume this is a guaranteed maximum price (GMP) construction project. One condition of the owner's selection was that you could not self-perform any work unless you received competitive bids and were the low bidder. What would have happened if the GC had been awarded this package to self-perform this work? Would the GC have asked for a change order from a negotiated client if they did not have the dollars in the estimate? Could the GC make a case for a change order proposal (COP) if they could cover the $50,000 from other savings in the estimate? Is this a good example of why a negotiated GC should be allowed to self-perform their normal work tasks?

_____
_____
_____
_____
_____

c. What can designers do to anticipate and mitigate these types of situations? Should design consultants package work according to standard subcontract or labor jurisdiction lines? Should owner-GC contracts make hard distinctions between the general contractor and subcontractors, and not use the term "contractor"?

_____
_____
_____
_____
_____

d. Who ultimately pays for these walls? Argue your case to the arbitration board and assume you are the PM for one of these companies. If it is not your fault, who is at fault:

   •   concrete subcontractor;
   •   general contractor;
   •   structural engineer;
   •   architect; or
   •   project owner?

_____
_____
_____
_____
_____

Case 52: Pulled quote

A metal decking supplier submitted a $300,000 bid to the negotiated general con-
tractor on a large distribution/warehouse facility. The second supplier's bid was
$600,000. After minor value engineering (VE) revisions, the general contractor
requested a subsequent bid from the low supplier, which was then even lower
than the original bid. After further strong-arm negotiations from the GC, the two
parties finally settled on a price of $250,000. The GC finalized their negotiations
with the client based upon this figure. One month later, the supplier declared that
they had a bid error and were pulling their bid. There was not a bid bond between
the general contractor and the supplier, and they had not yet signed their purchase
order (PO). The GC had negotiated and executed their contract with the client.
The general contractor made a plea to the client to be allowed to raise the bid by
$350,000 so that they could sign up the second bidder. The client denied. The
general contractor took this case to court. The lower court agreed with the client,
and ruled that: (a) the GC should have known the low bid was in error; and (b) the
GC should not have worked the low bidder's price down. On appeal, the upper
court reversed this decision and agreed with the general contractor's claim. Their
reasoning was:

1. There was no way for the general contractor to have known that maybe the
    higher bidder was in error.
2. The GC has proven bid error.
3. The client would have had to pay the additional $350,000 originally if the low
    bidder had either estimated correctly or not submitted a bid. The client was
    not due a windfall benefit.

How did the low supplier error in their post-bid actions? How did the GC
error? What could the owner have done to prevent this from happening? Do you
agree with the first court's ruling? Is the second court's ruling correct? What would
be the fairest way to resolve this situation for all parties?

_____

_____

_____

_____

_____

Case 53: Hostile project manager

You are an experienced project manager working for a flooring subcontractor.
Your younger counterpart on the general contractor's side is hostile to all of his
subcontractors. He challenges your RFIs, returning many without forwarding
them to the client or architect. He routinely cuts your change order pricing and
your pay request draws without consulting you. You decide to deal with this

individual by not bringing problems to his attention. You proceed "per plans and specifications," as he has directed you on numerous occasions. You have worked with the GC's chief executive officer (CEO) on previous occasions. He is a people person, and encourages cooperation and open communication at every opportunity. He has told you specifically that he is looking for long-term relationships with subcontractors who are part of the "team-build" solution. You have told him directly that his project manager is a problem, and the CEO has responded that he will look into it and appreciates your dedication to this project. The GC's project manager has found out that you have discussed this situation with his supervisor. Things are now worse than ever. Are you in the wrong? How do you deal with this situation now? How could you have dealt with it differently from day one? Once you have gone up the ladder and over someone's head with a problem, can you ever go back down for a solution?

_____

_____

_____

_____

_____

Case 54: Fire protection heads

You are within two weeks of turnover of a three-story build-to-suit office building project. Many of the city inspections are complete and it appears that a certificate of occupancy (C of O) will be obtained on time. Only the fire protection, elevator, and life safety inspections remain. The mechanical, electrical, fire protection and life safety were all design-build systems. During the punch list, it is discovered that the fire protection subcontractor has installed a different type of sprinkler head on each floor. The first floor uses concealed heads with white escutcheons to match the ceiling tiles. The second floor uses decorative chrome heads. The subcontractor installed semi-recessed heads on the third floor. This was not picked up prior as a person cannot physically see more than one floor at a time. The client is very upset. Can you require the subcontractor to change them all out now? Would this be a good move given the status of the inspections? Who is responsible to check and approve the shop drawings and submittals for design-build subcontractors? What should the general contractor have done to prevent this from happening? What can you do now?

_____

_____

_____

_____

_____

7

START-UP

Including preconstruction, mobilization, and value engineering

Introduction

Project start-up is a subset of preconstruction planning. Once the general contractor (GC) has been notified that the project has been won, the project team, headed by the project manager and superintendent, must plan project start-up activities. Efficient project start-up activities are dependent on a good start-up plan. Both the project manager and the superintendent should participate in the preconstruction conference. This meeting is conducted either by the owner or the designer to introduce project participants and to discuss project issues and management procedures. The notice to proceed (NTP) is often provided at the preconstruction conference, along with verification of owner financing and needed permits. The NTP authorizes the contractor to start work on the project, and once received the project manager should buy out the project by awarding subcontracts and ordering needed materials and equipment, especially those long-lead items.

Start-up activities involve establishment and organization of the project office and staff, construction site layout, and mobilization of the jobsite. The project manager selects the project management team and involves them in start-up planning. Paper mobilization, including contract files, project office administrative procedures, and correspondence management systems, is also established. Award of subcontracts, procurement of materials, and actual mobilization must wait until the NTP is received, but the planning should be completed prior to its receipt, so implementation can begin immediately upon receipt of the notice.

A schedule of submittals and a schedule of values are also prepared and submitted to the owner or owner's representative for approval during project start-up. Submittals will be discussed in Chapter 8 and the pay request schedule of values is discussed in Chapter 9. Project-specific quality control and safety plans need to be finalized, and, if required by contract, submitted to the owner or owner's representative. Case studies highlighting quality control and safety planning are discussed in Chapter 11.

The actual physical mobilization onto the jobsite is the responsibility of the project superintendent, but there may be strategies regarding exactly when to mobilize to assure reimbursement for negotiated preconstruction services, as discussed in cases in this chapter. This chapter includes examples of projects that struggled with reimbursement of mobilization and preconstruction services. Physical mobilization incudes:

- initial surveys, including property corners (by the project owner) and building corners and grid lines and site utilities (by the GC);
- site camp layout, including trailers and hoisting requirements;
- temporary erosion control, including storm water retention ponds;
- access roads and parking; and
- site fencing and signage.

Negotiated projects may include a preconstruction phase where the general contractor is selected early and assists the design team and project owner with a variety of tasks. It is important that there is a clear understanding of work to be performed, the amount of the preconstruction fee, and commitments for a construction contract, if appropriate. Some of those preconstruction services include:

- estimating;
- scheduling;
- constructability reviews;
- quality control and safety plans;
- identification of long-lead materials; and
- value engineering.

Case studies

Case studies in this chapter include:

55. No preconstruction agreement
56. Missed preconstruction meeting
57. Early mobilization
58. Formal value engineering
59. Private value engineering

Most of these case studies overlap with other primary topics, and other case studies throughout this book also involve start-up.

Toolbox quotes

Mobilize early, but not too early.

Day one of the contracted construction schedule is predicated upon receipt of a signed and executed contract, proof of owner financing, issuance of building permits, and notice to proceed.

Case 55: No preconstruction agreement

You are the project manager for a general construction firm and you think that you have successfully sold your services to a client who is building a golf course clubhouse. You have been performing informal preconstruction services for three months. This includes budgets, schedules, value engineering, meetings, meeting notes, and constructability analysis. You did not have a formal signed agreement for your preconstruction services, as depicted in Figure 7.1. You later discover that the designer has a contract to complete 100% plans and specifications. You also discover that the client intends to competitively bid the project out. It is rumored that your client may have been receiving preconstruction input from another general contractor in parallel with your efforts. You have spent $20,000 already out of pocket on preconstruction services. Did you make an error, and if so how? What do you do now? Why do GCs offer preconstruction services below cost and sometimes even for free?

Case 56: Missed preconstruction meeting

A utility subcontractor receives a permit to connect a new branch water line to a main in a heavily traveled street. The subcontractor is working for a general contractor. They shut the line down at 7:00 a.m. and begin tearing up the street. The city shows up at 9:00 a.m. and shuts the project down because the contractor did not hold the prescribed preconstruction conference. The street has now been excavated to the centerline, a six-foot-deep shored trench exists, and the water main has been exposed. Obviously, the error is the contractor's (but which one?), and the answer to the question "What did they do wrong?" is easy. But now, is the city correct in shutting the project down? Morning rush hour traffic was obviously impacted but is now over. The afternoon rush

---

Preconstruction Contract

This agreement is made the 5th day of January 2019 between:

The Project Owner: Physicians and Associates, LLP
2242 Second Ave
Rainier, WA 98111

And the Preconstruction Contractor: King County Builders, Inc.
4100 SW Hilltop Road
Rainier, WA 98111

For the following Project: Health and Wellness Medical Office Building

The Project Owner and Preconstruction Contractor agree as follows:

1. That during the development of the design and prior to the start of construction, the contractor will provide preconstruction services as follows:
    a. Attend weekly coordination meetings chaired by the Architect
    b. Prepare two complete budget estimates at the completion of schematic design and design development documents
    c. Develop cost analyses of design options as required
    d. Conduct value engineering studies as necessary to achieve project budget goals
    e. Meet with consultants to assist with development of design
    f. Generate bidding interest among the subcontractor and supplier industry
    g. Conduct constructability reviews at completion of each design submission and as requested by the design and project owner teams
2. Prepare a complete guaranteed maximum price estimate and be prepared to enter into an AIA A102 construction contract at the 90% construction document stage assuming acceptance of the estimate and contract terms by the project owner
3. Preconstruction services will be billed at the following rates, including all markups and burden, for a not-to-exceed amount of $375,000.00:
    a. Project manager: $120/hour
    b. Project superintendent: $110/hour
    c. Estimator and scheduler: $100/hour
    d. Project engineer: $ 85/hour
    e. Materials invoiced at cost plus 7% markup
4. Invoices will be submitted at the end of each month, including all necessary backup and subject to audit by the owner's accountants, and, if approved will be paid by the 10th of the following month
5. This is assumed to be a complete agreement and no other obligations or commitments are implied unless explicitly stated or added by mutually approved change orde r.

Approved by:

King County Builders, Inc.   Physicians and Associates, LLP

Jack Adams   *David O'Neal*
Jack Adams   David O'Neal
Project Manager   Owner's Representative

FIGURE 7.1 Preconstruction contract

hour traffic will begin at approximately 3:00 p.m. The continued shutdown of the water line will impact fire service and water service to the local neighborhood. What should the course of action be? Should the general contractor or subcontractor be fined? Should the general contractor or subcontractor be eliminated from bidding on future projects? In the public arena, can a contractor be eliminated from bid lists? Whose responsibility was it to schedule the preconstruction meeting?

Case 57: Early mobilization

A general contractor had previously submitted a budget of $14 million to the owner of a medical office building (MOB) project. The budget was based on 70% design documents. The negotiated contractor was selected based on the experience of its project team and its approach to the project. The contractor had been given a preconstruction services contract for $20,000 to join the owner's project delivery team. The contractor participated during the balance of the design phase of the project, performing value engineering and constructability analysis, and providing input to the construction drawings. During the last two weeks of the preconstruction phase, the general contractor mobilized on the jobsite. The owner neither directed nor stopped them from doing so. The terms of the prime construction contract had not yet been finalized. The contractor set up the site camp, brought temporary utilities to the site, and began the initial surveys and layout. The construction drawings were issued and incorporated all of the team's input, along with the city permit comments. The general contractor then prepared a $15 million guaranteed maximum price (GMP) estimate based upon these revised documents, which exceeded the previous budget by $1 million. The owner was extremely upset and would not listen to explanations or reasoning why the estimated costs had increased. The general contractor was asked to move off of the site. When the GC requested to recover the additional $15,000 they had incurred for the two weeks on the site, the owner refused payment. Should the general contractor have mobilized onto the site? Why would they have been motivated to mobilize without a contract? Does the GC have any recourse for payment? What steps would you suggest for both the owner and the GC now? If the GC was entering into a lump sum contract, would their actions have been different, and why?

Case 58: Formal value engineering

Many public education construction projects are required by this state to conduct formal value engineering (VE) studies. This occurs early in the design development phase. It lasts approximately one to two weeks and employs several outside consultants, none of which will ever work on the project. The cost of this study may be in the $50,000 to $60,000 range. Most of the ideas put forth by the team are totally without merit. Many of the VE ideas are for just a few hundred dollars. The VE proposals are included in a log very similar to a change order proposal log to be utilized later in the project (see Table 7.1). Rarely are any of the proposals accepted into the design, yet all of the owner and designer team members are satisfied that they complied with the intent of the law. Is this true value engineering? Is it fair to the taxpayers? What should be the prescribed process? Should a percentage of the construction budget be required to be either proposed as value engineering ideas and another percent accepted into the design? Should a third party employed by the state supervise the process? Should there be financial incentives?

Case 59: Private value engineering

Value engineering was a significant element of the preconstruction services provided by this general construction management team on a highly complicated medical facility. The negotiated GC competitively bid out all subcontracted areas, but the $66 million GMP was still 10% over the client's budget; $6 million would need to be saved through the VE process.

a.   A few of the VE proposals included the same scope but just a different method, material, or manufacturer, but most of the VE proposals involved elimination of scope, shelling areas, reduction of equipment redundancies, eliminating the owner's future expansion capabilities, and simply providing a product of lesser value or shortened life. Is this true VE?

TABLE 7.1 Value engineering log

King County Builders, Inc.

Health and Wellness, MOB

Value Engineering Log

Updated: 03/20/2019

Item	Description	Date Proposed	Value Proposed	Value Accepted	Value Rejected	Date	Comments
8	KCB install additional FIO equipment	02/10/2019	$25,000	$0	$25,000	02/12/2019	Owner will handle
10	Replace VCT with sheet vinyl floor	02/12/2019	$7,500	$7,500	0	02/12/2019	Submittal outstanding
11	Delete door vision glass	02/12/2019	($1,500)	0	($1,500)	02/15/2019	Safety issue
15	Use drought-resistant plants, delete irrigation	02/20/2019	($17,022)	($17,022)	0	02/25/2019	More sustainable

continued . . . .

b. Should the GC receive a portion of VE savings? Should there be some fee enhancement? Is there a potential for fee reduction? What incentive does the GC have to reduce the cost of the project?

_____

_____

_____

_____

_____

c. After the client has accepted the $6 million of savings, the design team must now undertake a significant document revision. Who pays for the redesign costs? If the architect provided the original $60 million target budget, does this change the scenario? Which documents are "contracted" to? Is this issue covered in standard AIA contract language?

_____

_____

_____

_____

_____

d. The GC and the low-bidding subcontractors proposed almost all of the savings ideas. The design team formally went on record disagreeing with many of the ideas, yet the owner still accepted them. Why would the design consultants do this? One proposal from the mechanical subcontractor involved a different exhaust system control valve. The valve never performed properly. The mechanical engineer did not help with analyzing the problem or the correction. The owner blamed the GC and the mechanical subcontractor who proposed this $150,000 savings, even though the owner accepted it. Are contractors liable for performance of VE proposals? Are they assuming design risk? If so, why would they propose any cost savings?

_____

_____

_____

_____

_____

e. Do contractors like VE? Do they give back a dollar for a dollar's worth of savings? VE can also be for more expensive items and increase the GMP, as shown in Table 7.1. Explain the term "life cycle costing." Do contractors inflate estimated costs when additive cost proposals are processed? Is VE a way for contractors to improve profits or cover for other estimate line item shortcomings?

_____

_____

_____

_____

_____

8

COMMUNICATIONS

Including RFIs and submittals

Introduction

"Communication" is a broad term that, loosely described, includes acquiring and transmitting different types of information. It is perhaps the most critical project management (PM) tool. A good technical project manager who knows how to estimate, plan, schedule, and execute a construction contract may fail, unless he or she also has good communication skills. Unless the project manager can communicate his or her needs, wants, and expectations, they likely may be unfulfilled. Any study of construction leadership shows communication skills as one of the strongest traits of both office and field construction leaders. Several formats and techniques have been developed to expedite the flow of information among members of the project team. There are many construction communication tools, including contracts, schedules, logs, and start-up documents. This chapter highlights case study examples of both good and bad uses of communications by contractors, designers, and project owners.

Construction communications may be written, oral, or electronic. Good communications are concise and focused to avoid misinterpretation. Communication documents and tools used by contractors include the following:

- Transmittals are used to forward other construction management documents and provide a written record of what was sent.
- Meeting notes are used to record the issues discussed and decisions reached at project meetings. They also list open action items and individuals tasked with resolving them. Meetings, especially weekly owner-architect-contractor (OAC) coordination meetings, are important communication opportunities for the project manager. Each meeting needs an agenda to provide focus and a note-taker to prepare the meeting notes. The exact format for the meeting

notes is not as important as the fact that notes were prepared and distributed to all attendees.

- Requests for information (RFIs) are used on the jobsite when the contractor is not sure of a detail or a dimension and requires supplemental interpretation from the design team. Depending upon the project size and the procurement method, there may be hundreds or even thousands of RFIs on a project, which warrant attention from all of the contracting parties.
- Submittals are a way for contractors to communicate to designers and project owners their understanding and intent of materials furnished, as well as proposed means and methods. Submittals are an active quality control tool, and the construction team's contribution to finishing the design and flushing out potential conflicts, similar to RFIs, as shown in Figure 8.1.

As stated at the beginning of this book, construction is a people business, and it is a large group of individuals working together that build buildings. The interplay between the owner's representative, architectural project manager, general contractor's project manager and superintendent, project engineers, and subcontractor office and field representatives relies on good communication processes. Several case studies in this chapter emphasize the importance of open and honest communications.

Case studies

Case studies in this chapter include:

60. RFI value
61. Dropped baton
62. Architect's administration
63. Three general contractor project managers
64. Two architectural project managers

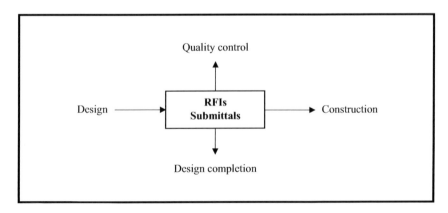

FIGURE 8.1 Active quality control

Most of these case studies overlap with other primary topics. Communication, to some extent, is involved in all aspects of construction management and in most of the cases presented in this book. Case studies 26, 54, 76, 78, 82, 83, and others also directly involve communications and document control.

Toolbox quote

If contractors would learn to communicate, many attorneys
would be out of a job.

Case 60: RFI value

a. A high-technology client and their architect engage in the design and construction of a multiphased multi-building research campus. The total design and construction process will take over 10 years. The buildings will all be similar in design and function, but not exact. The design is complex, and the first building will eventually cost approximately $500 per square foot. A general contractor (GC) is selected for building phase 1 after a competitive proposal. It is their hope to build out the entire campus. The general contractor's project manager and his project engineers generated 2,500 requests for information, and 500 change order proposals (COPs) were subsequently negotiated in this first phase. The contractor pointed to document discrepancies for most of the issues. In addition, there were over 100 submittals that were completely rejected by the design team. The GC and the subcontractors also pointed to the specification ambiguities for the submittal difficulty. There were not any outstanding claims, liens, or quality questions at phase 1 completion. The job was completed on schedule and there were not any safety incidents. All parties seemed to be amicable. This was a successful project, wasn't it?

---

b. Ultimately, the client engaged another general contractor for the second phase. The architect later boasted to the first general contractor's PM that his successor only wrote 1,000 RFIs and less than 200 COPs on phase 2. Very few submittals required complete rejection. What happened? Was the second GC team easier and more "client-friendly"? What did the design team do with all of the first GC's documentation?

---

c. The second GC continued on and built five more buildings and the remainder of the campus. The above RFI, COP, and submittal figures reportedly were reduced with each subsequent phase. Did the first project manager do a bad job? What does this say about RFIs and submittals with respect to the design and the close-out processes? Aren't RFIs and submittals actually part of the design completion and early "active" quality control? Who benefited from this process? Who paid for it? How could the first PM have kept his firm on the campus to construct the balance of the phases?

Case 61: Dropped baton

The GC you work for has just relocated you from one jobsite to another. The project engineer (PE) you are replacing was "relieved of his command." He was very experienced and had a very strong personality. You will be managing the work of the heating, ventilation, and air conditioning (HVAC) subcontractor, although your prior experience was all related to civil and structural work. You barely know the difference between pipe and duct. You find that your predecessor's desk is completely empty. You review the files for the HVAC subcontractor and find them very brief, disorganized, and unprofessional. No one at your new site has any knowledge of the work that your predecessor or the HVAC subcontractor were doing, other than they fished a lot together. You cannot find an executed copy of the subcontract agreement. The invoices and change orders for this subcontractor are not even close to tracking. Upon notifying the subcontractor that you will be handling their account, their PM indicates that he is very displeased about your predecessor's removal. "They had an understanding, and commitments were made." What do you do? What could be done to the general contractor's systems to prevent this "dropping of the baton"?

Case 62: Architect's administration

This project is an $8 million public elementary school remodel that was bid lump sum. There are numerous conflicts in the documents, many of which are associated with matching new to existing work. The project manager for the general construction firm has become frustrated with the lack of paperwork from the architect. The architect appears to have run out of construction administration funds. Some of the problems and responses to requests for support are listed below:

- Responses to the RFI process include, "I don't want a written question, just give me a telephone call."
- The architect will answer written RFIs with a verbal response.
- She will not meet in the field and review actual discrepant conditions.
- Written responses quite often just indicate "see the plans or specifications."
- Submittals are returned late, and they often do not include any disposition.
- The architect misses many weekly construction meetings, showing up at some late and leaving others early.
- She never brings her meeting notes and does not acknowledge ever receiving them.
- She is also not reviewing change orders in a prompt fashion.

What should the GC's project manager do to resolve this issue? What could the school district have done to prevent this from happening? What risks do both the general contractor and the school incur if this situation is left unchecked? What risks is the architectural firm exposed to?

_____

_____

_____

_____

_____

Case 63: Three general contractor project managers

The construction market is very busy. This general contractor is having a difficult time obtaining good qualified project managers. The officer in charge (OIC) has hired three new project managers (A, B, and C) within the last few months. The OIC performed a diligent interview and reference check on each of these employees. She is just about to receive the following reports from her superintendents, who are working with these individuals on three separate projects.

Project manager A likes the detail. He is very thorough in his research and his work is very accurate. Unfortunately, all of this accuracy is taking too much time. He tends to hold RFIs and submittals and subcontractor change order requests too long. He has a tendency to hand out last week's meeting notes at the beginning of the next week's meeting. Delay and impact notices are now coming in from all of the subcontractors.

Project manager B has a field background. He was a journeyman carpenter before going to college to get a construction management degree. He spends most of his day outside of the trailer. He is prone to provide direction to the GC's and subcontractor's field forces, bypassing his superintendent and that of the subcontractors. His paperwork is being performed adequately during off-hours. The general contractor's superintendent is about ready to quit.

Project manager C was educated as a structural engineer. She has a professional engineering license and lets everyone know it. She is constantly second-guessing the design documents and design team direction, but she is always correct. She has written several negative letters to the design team, and is copying the city and the client with these letters. The design team is about ready to mutiny; even a subcontractor has written a letter requesting to be let out of their subcontract. Her project management paperwork is excellent and timely. She always returns a profit.

What should the officer in charge do? What will happen if these scenarios are left unchecked? If, due to a slowdown in the market: (a) who would be the first PM the OIC would let go; and/or (b) which one has the most potential to develop into a construction leader? Which type of PM are you?

Case 64: Two architectural project managers

As a project manager employed by a general contractor, you find yourself working with two different and difficult architectural project managers from the same firm on different projects. Your firm has a very good relationship with this architectural firm. Your OIC has told you to "deal with it" and not to damage your firm's reputation, and at the same time make a fair profit. Assuming that you cannot change these professionals' personalities and you have six months to go to finish each project, how do you deal with each of the individuals as described below?

a. Architect A is young and has a very strong personality. He develops and distributes his own copies of meeting notes. He publishes his own RFI, submittal, and change order proposal logs. He refuses to address your logs or notes in the meetings. His records consistently slant in his favor with respect to content, responsibility, and dates. He chairs any meeting he attends and demands that he sit at the head of the table. Change order proposals that you originate do not show up on his logs. He authored your contract and your AIA change orders. Many of your "issues" continue to be sidestepped.

_____

_____

_____

_____

_____

b. Architect B is pleasant to be around and very experienced, but appears at times to be laissez-faire. She appears to be tired and overworked. She does not take any notes during the meeting and continues to show up unprepared, forgetting her copies to important documents. She leaves meetings early for "prior commitments." She loses RFIs, submittals, and pay requests. She does not recall verbal or phone conversations, and does not acknowledge receipt of emails.

_____

_____

_____

_____

_____

9

PAY REQUESTS

Including liens, lien releases, and retention

Introduction

There are many important project management functions, as discussed throughout this book, but maybe the most important one is to be paid for their work. Project managers and superintendents may have all of the tools necessary to earn a profit on a job, but if the owner does not pay for the work, the contractor will not be able to realize a profit. Some subcontractor project managers do not acknowledge the importance of preparing prompt payment requests. If a payment request is not submitted on time, the general contractor (GC) will not likely get paid on time. Proper cash management skills are essential for the general contractor and its subcontractors; without it, they may find that they are unable to pay their material suppliers and craftsmen. Good cash management skills, just like good communications skills, are essential if one is to be a successful project manager.

Contract impacts

Pay requests are typically submitted once monthly first by the subcontractors to the GC and in turn from the GC to the project owner. The formats and times are specified in the special conditions of the contract. Regardless of the type of contract, many of the procedures are similar, but there are some pay request differences associated with different types of contracts.

Payment on a *lump sum contract* is based on a schedule of values (SOV) and percentage completion. Front-loading and overbilling can occur on this type of contract. General contractor records are rarely audited in a lump sum contract. A *unit price contract* allows for payment based upon quantities actually installed.

If the contractor is to be paid $2,000 per ton for structural steel installed, and has installed 50 tons, then they will be paid $100,000, less any agreed upon retention. This process is quite objective and can be facilitated by an outside quantity measurement individual, firm, or team. Payment on a *time and materials* (T&M) contract is based on actual labor hours times a contract labor rate plus reimbursement for materials based on supplier invoices. Similar to estimates, *guaranteed maximum price* (GMP) contracts are a hybrid of lump sum and T&M. GMP contracts also include a pre-established SOV and may also have line items that are invoiced on an actual cost basis.

Receipt of timely payment is one of the most important responsibilities of the project manager. The exact format for submitting pay requests will vary depending on the type of contract. A schedule of values is used to support payment applications on both lump sum and GMP contracts. The project manager is responsible to develop the payment request, make sure payment is received, and subsequently see that his or her subcontractors and suppliers are paid. If payment has not been received on time, the project manager should contact the owner to determine the cause. The financial relationship with the owner is the project manager's responsibility. The same scenario holds true with respect to subcontractors and suppliers. The project manager must ensure that they are also paid promptly.

Owners may withhold retention, which is a portion of each payment to ensure timely completion of the project. The retention rate is also specified in the special conditions of the contract, and often amounts to 5% or 10%. A monthly invoice summary page that includes a 5% reduction for retention and a 10% addition for state sales tax follows as Figure 9.1. Liens can be placed on a project if subcontractors or suppliers are not paid for their labor or materials. To preclude liens, owners require lien releases with payment applications. Understanding liens and lien releases and retention are crucial project success measures for contractors, as shown in several of the following case studies.

Case studies

Case studies in this chapter include:

65. First-time developer
66. Subcontractor overbilled
67. Supplier's lien
68. Shoring retention
69. Which site to lien?

Most of these case studies overlap with other primary topics. Case studies 26, 48, 49, 70, 79, 94, and others also involve pay requests.

PAY REQUEST SUMMARY
Where the basis of payment is a Guaranteed Maximum Price

Client: Physicians and Associates, LLP
Address: 2242 Second Ave., Rainier, WA 98111
         (206) 527-4424

Contractor: King County Builders, Inc.
Address: 4100 SW Hilltop Road, Rainier, WA 98111
         (206) 527-4343

Project:	Health and Wellness, MOB	Contractor's Project #	869
Billing Period: From:	11/01/19	Contractor's Request #	8
To:	11/30/19	Contractor's Invoice #	402

Original Contract Price:			$14,886,216
Approved Change Orders:	1 through 4		$752,119
Current Contract Price:			$15,638,335

Total work completed to-date: (See attached detailed SOV)		$10,022,045
Less total retention held at a rate of:	5%	-$501,102
Plus applicable State Sales tax added at a rate of:	9.00%	$901,984
Subtotal completed to-date less retention, inclusive of Sales Tax:		$10,422,927
Less prior applications for payment, net of retention and including tax:		$8,995,520

Net amount due this pay period: **$1,427,407**

Contractor's Certification:
I hereby do certify that to the best of my knowledge the above accounts reflect a true and accurate estimate of the values of work completed and that all amounts paid prior have been distributed according to existing contractual requirements and the laws in existence at this jurisdiction at the time of contract.

Contractor: *Jack Adams*   Date: 11/30/2019

Architect's Certification:
In accordance with the contract documents and limited on-site evaluations we have no reason to believe that the accounts presented by the contractor are not reflective of the work accomplished to date.

Architect: *David O'Neal*   Date: 12/01/2019

FIGURE 9.1 Pay request summary

Toolbox quotes

What is the most important thing a contractor does?

The project manager is the ultimate accountable party.

Sell the pay request.

Cash is an important project management tool.

Case 65: First-time developer

This real estate broker was envious of real estate developers and the "reported" large profits they make. He decided to try his hand in this area, although he did not have any design or construction experience. With insider knowledge, he came across a piece of property that was available from another developer which had a design and permit in hand for a planned 20-unit condominium complex. The first-time developer dismissed the architect and acted as his own owner's representative shortly after he closed escrow on the property.

The deal also came with a general contractor who had worked on preconstruction for the previous developer. This new developer negotiated a $20 million lump sum construction agreement with the same general contractor. The contractor authored the contract and proceeded with the project. The contractor was actually more of a construction manager than a general contractor, as is often the case with residential work. They subcontracted out all of the work and did not have a full-time superintendent on the project. The new developer was not introduced to any of the subcontractors and he did not have a written subcontractor list. The general contractor and the subcontractors did not post any bonds.

The residential market was good and the developer had 80% of the condominiums pre-sold (cash pending completion), even though the project was just drying in (roof on, siding on, windows in) and ready for mechanical and electrical rough-in. All relations had been proceeding fine with the contractor. The quality of the work was adequate, there were very few change orders, and there had not been any safety incidents. The construction schedule was only 12 months long, and the contractor was on, or even ahead of, schedule.

The new developer had obtained a loan for all of the construction costs and leveraged other personal holdings against the land. Progress payments of $12 million had been paid on time. The pay request schedule of values was only five items long. There were not any conditional or unconditional lien releases submitted by the subcontractors or the general contractor. Retention was not being withheld from the GC. It turns out the owner of the construction firm was having an extra-marital affair with his treasurer. For the entire course of the project, she had only paid about half of the subcontractor invoices. Without any notice, the two of them skipped town and disappeared to the Caribbean, and were never heard from again.

The subcontractors all immediately demobilized and filed liens against the property. They had all properly filed their materialmen's notices before the project began, but the new developer was inexperienced in lien management and did not retain any of these notices. None of the subcontractors would return until the developer had paid them their outstanding shares of the work in place, which was now approximately $6 million. The bank begins to pressure the developer. Many of the buyers begin to look for other residences. Some begin legal proceedings against the developer. Now what? List at least 10 project management errors this first-time developer made.

_____
_____
_____
_____
_____

Case 66: Subcontractor overbilled

After a non-performing drywall subcontractor has been terminated, it is discovered that they had overbilled your general construction firm, and you, as the project engineer, had processed their invoices. Their contract was for $1 million. The subcontractor billed, and you authorized payment of $750,000. You had originally received competitive bids and the subcontractor was barely low. The drywall subcontractor had been submitting a 10-line item schedule of values with the pay request, and you were sharing it with your superintendent to verify level of completeness. You now bid out the remaining work, and receive very competitive bids, all near $500,000, to finish the work. You are $250,000 short. Your officer in charge jumps down your throat for allowing this overbilling. You have been very busy, managing 10 subcontractors on this job. How did this happen? Where do you suppose the PM was during this process? Who is at fault? What can be done next time to assure it doesn't happen again? How do you solve this specific problem? Can your employer garner your wages?

_____
_____
_____
_____
_____

Case 67: Supplier's lien

You feel that you practice very proactive lien prevention procedures. On this project, you received materialmen's notices from your fire protection subcontractor. Each month, you received proper conditional lien releases. At the completion of the job, you received an unconditional lien release and exchanged it for the fire protection subcontractor's retention check. One month later, you received word that a lien had been filed by a fourth-tier piping supplier from another state for material they supplied to a piping fabricator for your subcontractor. (_Note:_ many different states have different rules with respect to lien laws.) You did not know this supplier existed. You paid the fire protection subcontractor, the subcontractor paid their pipe fabricator, but the fabricator did not pay the pipe supplier. The amount in question is $40,000. How did this happen? How can it be resolved most easily? What is the correct and legal resolution? Is the lien valid? Can the supplier legally remove the pipe? What does the client do now? How are liens removed? What should you do in the future to prevent this from happening?

_____
_____
_____
_____
_____

Case 68: Shoring retention

This GC hired a subcontractor to install a conventional temporary shoring wall for an underground parking garage. The subcontract agreement has an inclusion that calls for the subcontractor's retention to be released within 30 days after fulfilling all of its contractual requirements. The GC's contract with the owner only allows early release of retention with the owner's prior approval. One month after all shoring work was completed, the GC submits the normal pay application to the owner, but includes a request to release all of the shoring subcontractor's retention. The owner refuses to release retention for the shoring wall until the parking garage structural work is complete (which will take another 60 days), and is holding up the entire pay request. What should the GC do? Should the owner release the retention? If the owner continues to refuse, what recourse does the subcontractor have with the GC? What recourse does the shoring subcontractor have with the owner? If they lien the property, should the GC pay the subcontractor out of its own funds? What position does the construction lender take on early retention release? Is early retention release (for select subcontractors) a good practice (for any or all parties), and why? What should have been done to have prevented this situation from occurring?

_____
_____
_____
_____
_____

Case 69: Which site to lien?

This client and general contractor were under contract for two entirely different projects on two different sites. Some, but not all, of the subcontractors were the same. The work on the first project appeared to have been completed according to expectations. Six months after the retention was released on the first project, the owner discovers numerous cracks in the concrete and the repairs appear to be extensive. The exact cause is unknown and will take months to determine. In the meantime, the owner holds back any further payments to the GC on the second project. Can they do this? The subcontractors and general contractor will lien, but which property do they lien against? If the owner cannot withhold funds, what recourse do they have to remedy the first project? Can the contractors stop work on the second project? Is this wise?

10

COST CONTROL

Including lean construction

Introduction

General contractor and subcontractor project managers and superintendents are responsible for ensuring their respective projects are completed within the time allowed by the contract and within the project budget. The popular phrase "under budget and ahead of schedule" is easier said than done. To accomplish this challenging task, cost and time controls are established to monitor progress throughout the duration of the project. The project team wants to ensure that all contractual requirements are completed as early as possible while earning a fair profit. Early completion of a project results in reduced jobsite overhead costs, enhancing potential profits. Case studies in this chapter will examine both good and bad techniques for cost control.

Estimate development is the first phase in the cost control cycle, as depicted in Figure 10.1. Cost control begins at that time by assigning cost codes to the elements of work identified during the work breakdown phase when developing the estimate. After a project has been awarded, the project team must correct any errors, buy out subcontractors, and input the estimate into the cost control system. Work package analysis provides the project manager and the superintendent a method for tracking direct labor cost, which represents the greatest risk on the project.

Cost codes allow the project manager and superintendent to accurately record and monitor project costs and compare them to the estimated costs. The objective is not that the team has to rigidly keep the cost of each element of work under its estimated value, but to ensure that the total cost of the completed project is within budget. This is analogous to driving on the road. You don't have to exactly follow the "line down the middle of the road," but you need to stay on the road. Another more advanced

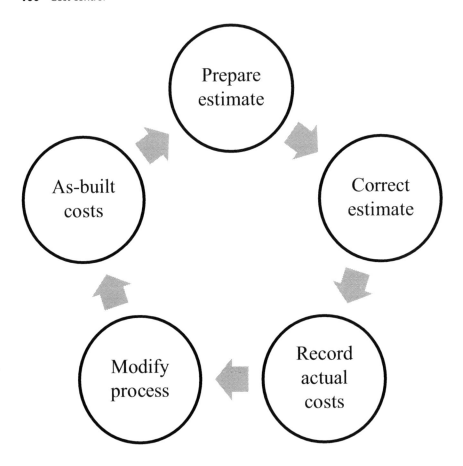

Where does a circle start?

Where does a circle end?

FIGURE 10.1   Cost control cycle

technique for cost and time control is earned value analysis, which compares the budgeted value of work completed with the actual cost incurred. Lean cost control techniques have been borrowed from production industries such as automobile manufacturing and applied to construction. Lean construction case studies new to this edition have also been included in this chapter. As-built estimates are developed to create unit price data that can be used on future estimates. Development of the as-built estimate is the last step in the cost control cycle, which leads into an improved database and the next project estimate.

Case studies

Case studies in this chapter include:

70. General contractor cash flow
71. Cost overruns
72. Low forecast
73. Residential developer
74. Lean GC
75. Lean vice president

Most of these case studies overlap with other primary topics. Case studies 9, 27, 32, 33, and others also involve cost control.

Toolbox quotes

Can we really "control" costs, or do we provide the project team with the appropriate tools to "manage" costs?

Does the office share the estimate with the field? They should!

Cost coding: garbage in, garbage out.

Case 70: General contractor cash flow

You are a new project manager (PM) for a small general construction (GC) firm whose annual volume is $100 million. All of the work is proceeding well on your first project, which is a local retail facility. You have been preparing the pay requests and picking up the checks from your client. You know that your construction firm is being paid on your project. You have been approving the subcontractor and supplier invoices. You begin to hear rumors that your company may be in financial difficulty, which is not because of you or your project. Your subcontractors have been complaining that they are not getting paid. When you check with your comptroller, she indicates that the subcontractors' checks are "in the mail." Your client also begins to hear complaints from your subcontractors, and pre-lien notices are being filed. What do you do?

Case 71: Cost overruns

a. Three months into a lump sum project, this project manager has realized that his company is overrunning costs on half of the direct work activities. They are under-running on the other half. The bottom line looks okay, maybe even on the plus side. Is this possible? Can he do anything about the codes that are overrunning? Since the bottom line looks okay, should he even worry about it? This project's superintendent does not want to report any single line item cost overruns to the home office, and asks the PM to forecast the original estimate in each case so that the variances zero out. How can the PM cover up this situation? What are the ramifications if he does what the superintendent is asking? As long as the project makes the original fee, does it really matter how individual cost codes turn out?

_____

_____

_____

_____

_____

b. This is a similar situation to the above, except all of the codes appear to be overrunning their estimates. The project looks like it is going in the tank to a tune of approximately $500,000. What are some of the reasons this overrun could be occurring? Should the project manager and the superintendent just hold on and ride it out? What are some potential resolutions you could recommend to the jobsite team? Should they ask for assistance from the home office? Should they start looking for a new employer?

_____

_____

_____

_____

_____

Case 72: Low forecast

A general contractor's project manager negotiated a cost–plus–fixed-fee (CPFF) contract with an apartment developer but without a guaranteed maximum price (GMP). The contractor's scope was limited to utilities, site work, and foundations for this hillside project. There were extensive concrete foundation walls and shoring systems required in the design in order to place all eight wood-frame buildings on this steep site. The superintendent and project manager calculated the amount of fill necessary to place behind the retaining walls and stored that fill on the site during excavation and concrete work. The balance of the excavation was exported from the site. The dirt was clean structural fill and the earthwork subcontractor likely sold it for a profit. All was proceeding well with the project. The developer asked the GC's project manager to share his monthly forecast against the original $5 million budget. At that time, the PM anticipated an under-run of $100,000 compared to the original budget. The developer then used this forecasted $100,000 savings to upgrade the apartment kitchen equipment specifications.

It turns out that the construction team underestimated the amounts of structural fill necessary for the walls and had to purchase select fill to be brought back to the site. The anticipated budget under-run was eaten up by this extra dirt. The developer claimed the contractor was negligent by hauling off needed materials earlier in the project and the contractor was liable for the $100,000 they had now spent. Is the developer correct? Should a PM share the forecast with a client? Is he or she required to contractually share the cost forecast on an open-book job? Will this cause the PM to be more conservative with cost forecasts in the future?

Case 73: Residential developer

A speculative residential sole proprietor builder/developer began his career with one to two houses at a time. He gradually worked up to 5- and 10-lot mini-developments. He had a good product and was pricing his homes competitively. The builder had established a good reputation with his banker, realtor, and the subcontractor industry. He had always made his payments on time. He had an opportunity to step into a larger project and he purchased a 50-lot subdivision. The land was priced 20% too high, but the market was hot and he figured he could make a big killing. Most of the land cost was leveraged. He built and sold five houses without a problem. His realtor was encouraging him to step up production and keep the prices high. He decided to proceed with the next 20 all at the same time, which was not his normal business model.

About halfway through framing, interest rates jumped two points and the residential market came to a screeching halt. The developer continued with the 20 houses under construction but was only able to move two of them, and they went for 10% below his asking price. The remaining 18 homes sat for six months with no activity. The realtor was now suggesting the developer cut his prices dramatically. He dropped his price to below cost and was able to sell eight more. The last 10 were held for another six months. All of his creditors were filing liens. The bank had made several extensions on the land and construction loans, and was now threatening foreclosure. The developer was able to sell the remaining 25 unimproved lots to another developer, but they went for 30% less than the original purchase cost, which ultimately put him further in debt without any possibility for recovery.

This first-time developer filed for bankruptcy. His reputation was tarnished in the small community his family lived in, and they eventually moved to another state, attempting to run from criticism and the creditors. How did the developer error? What mistakes did the subcontractors make? Should the subcontractors have put community and relations first and held off on the liens? Should the bank have loaned him for the original land purchase? Should the bank also have held off on foreclosing? What should the bank have done to prevent the developer from becoming so extended? Was the realtor at fault?

_____

_____

_____

_____

_____

Case 74: Lean GC

Lean construction techniques have been adapted for construction in the last 20 years from the automobile industry. Lean basically means to build construction projects as cost-effectively as possible and minimize waste. Adopting a lean approach to construction has been received well by some contractors and resisted by others. This case study private developer has recently attended a lean presentation by a college professor and is on the fence whether to require her open-book time and materials (T&M) project to be built by a lean contractor advocate. She is interviewing two perspective general contractors.

a. GC1: Argue why your company has always built in an economical fashion, even it if was not labeled as "lean." Your construction firm has been successful in securing both competitively bid and negotiated projects and is returning a fair profit. Use concepts such as preconstruction services, constructability reviews, value engineering (VE), total quality management (TQM), vetting best-value subcontractors, cost and schedule controls, foremen work packages, earned value, zero punch list, zero time loss due to safety accidents, and others to pitch your company to this client.

_____

_____

_____

_____

_____

b. GC2: Your firm has sent all of your superintendents, project engineers (PEs), and project managers to Associated General Contractors of America (AGC)-sponsored lean training classes. Everyone is lean-certified, and your business cards and letterhead have been modified accordingly. Your company's goal is to be the leanest contractor in your state. Prepare an argument why your firm is now more cost-effective than your competition and you are able to validate that lean construction projects are finished less expensively, with less waste than those that are built by standard non-lean GCs.

_____
_____
_____
_____
_____

Case 75: Lean vice president

This large design–build mechanical subcontracting company performs heating, ventilation, and air conditioning (HVAC), processing piping, plumbing, building controls, and fire protection all in-house. If they were a general contractor, they would be the third largest in this metropolitan area. The construction company employs the top mechanical engineers and has 200 union mechanics in their fabrication shop. They sell fabricated ductwork and piping to many of their smaller competitors.

The HVAC vice president (VP) has just attended a week-long lean construction technique seminar that included topics such as activity-based costing (ABC), just-in-time (JIT) planning, pull planning, efficient jobsite material laydown and handling, and supply chain material management. The VP is tasked with transforming her company into the top lean mechanical contractor in their three-state region. She has hired you as an expert lean consultant to help her phase in this transformation.

You are to prepare three lists (A, B, and C) of five actions each you recommend she implement. Each of these plans includes a lean workshop hopefully conducted by you. The "A" list includes easy minor changes that do not require buy-in or approval from any of her other team members. The "B" list of five improvements involves a pitch to the project managers, superintendents, and project engineers. They will have to devote some of their time and energy to incorporate these suggestions into their project plans. The final "C" list will have cost and resource commitments needed from the subcontractor's board of directors (BOD). They will be the hardest to sell and will require the most time to implement. Consultants usually bill contractors on an hourly basis, so make your pitch stick!

_____
_____
_____
_____
_____

11
QUALITY AND SAFETY CONTROLS

Quality control

Project controls were customarily represented in academia with a triangle and included cost, schedule, and quality. The triangle is of course a very strong geometric shape, but in our project management book we added a fourth side to the triangle – for safety – to create the controls diamond. A project that is completed within budget, on schedule, and with acceptable quality is a failure if someone is seriously injured. The diamond is also a very strong shape. All four of these critical project attributes require documentation and clear communications by the project participants, as discussed throughout this book. Document control has been added here as a fifth dimension, and is reflected in the controls house in Figure 11.1. Your house, or home, is also a very strong structure.

Quality control is one of the five project control functions we have been discussing throughout this book. The quality of construction has short-term implications affecting material and labor costs on a project and long-term implications affecting the overall reputation of the construction firm. The firm's greatest marketing assets are satisfied owners, and delivering quality projects is critical to achieving customer satisfaction. Example case studies in this chapter unfortunately reflect a couple of contractors whose focus was more on reducing costs than maintaining quality, much to the disappointment of their clients. The project manager (PM) and superintendent must work together to ensure that all materials used and all work performed on a project conform to the requirements of the contract plans and specifications. Nonconforming materials and work must be replaced at the contractor's cost, both in terms of time and money. This means that the contractor must bear the financial cost of tearing out and replacing the nonconforming work, and that additional contract time is not granted for the impact the rework has on the overall construction schedule. Attention to quality control is essential

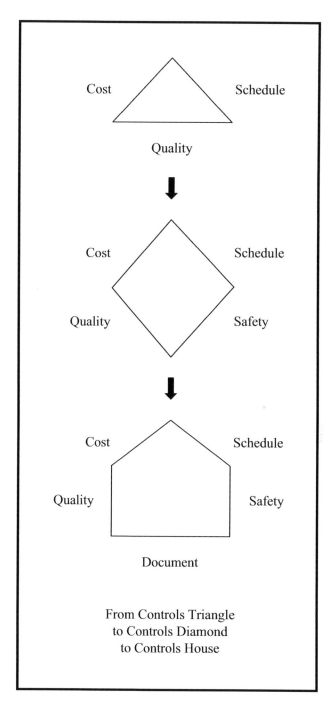

FIGURE 11.1 Evolution of project controls

to ensure all contract requirements are achieved with a minimum of rework. This will only be achieved through a proactive quality focus from all members of the project team, from the corporate offices all the way through to the subcontractors and foremen.

Project-specific quality control (QC) plans are developed during the pre-construction phase or project start-up to document the QC organization and procedures to be used on the project. Some project owners on negotiated contracts require project QC plans to be submitted with cost and schedule proposals as well as project-specific safety plans. Other owners may require that QC plans be submitted at the preconstruction meeting. The exact procedures to be used and the organization of the QC team are identified in the project-specific quality control plan. This is a prime example of "active" quality control versus "passive" quality control, taking measures to prevent mistakes and assure the project goes well the first time around compared to repairing bad work after the fact.

For any contractor to achieve project success, both in terms of a satisfied owner and a profitable project, implementation of an effective quality management program is essential. Poor-quality work costs the contractor both time and money, and can cause the loss of future projects from the owner. The project manager and superintendent must ensure that all materials and work conform to contract requirements. The objective of a quality management program is to achieve required quality standards with a minimum of rework. Quality materials must be procured and qualified craftsmen selected to install them. Workmanship must meet or exceed contractual requirements.

Quality control inspectors may be specified on large contracts or for complex projects. Some contractors attempt to substitute a foreman or assistant superintendent who already has other responsibilities. If a full-time inspector is specified, the general contractor must provide one and include the cost in the general conditions part of the estimate. On smaller projects, the quality control and safety inspection functions may be combined and/or performed by field supervisors with other responsibilities.

Quality control inspectors, whether they be a contracted third party or a part of the GC's internal staffing, should be used to inspect all work, whether performed by the contractor's workforce or a subcontractor's. Materials are inspected for conformance with contract specifications upon arrival on the jobsite. Mock-ups may be required to establish workmanship standards. Each phase of work is inspected as it progresses, and a pre-final inspection is conducted when most of the work has been completed. All deficiencies identified during the pre-final inspection are recorded on the punch list. When all outstanding deficiencies have been corrected, the final inspection is conducted.

Safety control

Similar to estimating and scheduling, there are several good textbooks on safety, including *Construction Project Safety* by Schaufelberger and Lin. Most project management books, such as *Management of Construction Projects: A Constructor's Perspective* by Schaufelberger and Holm and *Introduction to Construction Project Engineering* by Migliaccio and Holm, include at least one chapter dedicated each to quality control and safety control. What follows here is a very brief introduction on safety controls. Construction is one of the most dangerous occupations in the United States, accounting for about 10% of the disability injuries and 20% of the fatalities that occur in the industrialized workforce. There are two major aspects of project site safety: safety of persons working on the site and safety of the general public who may be near the project site. Both aspects must be addressed when developing project-specific safety plans.

The Occupational Safety and Health Administration (OSHA) has the primary responsibility for establishing safety standards and enforcing them through inspection of construction worksites. The Occupational Safety and Health Act, which established OSHA, contains a provision which allows states that desire to administer their own industrial safety programs to do so, as long as their requirements are at least as stringent as those imposed by OSHA. About half of the states have opted to administer their own programs.

Successful contractors have all recognized the importance of safety management and have developed effective company safety programs that include new employee orientation, safety training, and jobsite safety surveillance. The effectiveness of these programs, however, is directly related to management's commitment to safety. Project managers and superintendents are responsible for the safety of workers, equipment, materials, and the general public on their project sites. They must set the standard regarding safety on their projects and enforce safety standards at all times. Jobsite safety is a significant project management issue. Creating a safe working environment is a function of the physical conditions of the working environment and the behavior or working attitude of individuals working on the site. A safe working environment results in increased worker productivity and reduces the risk of injury. Accidents are costly, leading to disruption of the construction schedule, and require significant management time for investigation and reporting.

Potential safety hazards exist on all construction sites, and mitigation measures are needed to minimize the potential for injury. Most construction projects are unique, and construction workers are constantly expected to familiarize themselves with new working environments. In addition, craftworkers may only work on a project site during certain phases of work and then move onto another project site. This continuing change in the composition of the workforce presents significant changes to the project team. Another significant safety challenge is the increased employment of workers for whom English is a second language. These workers often have difficulty reading and understanding safety signage.

The key to a proactive safety control program is accident prevention. This is accomplished by identifying all the hazards that are associated with each work activity and developing plans for eliminating, reducing, or responding to these hazards.

This is known as job hazard analysis. Substance abuse can also result in accidents, so an effective substance abuse program is needed to remove individuals from the jobsite who are under the influence of drugs or alcohol. Personal protective equipment (PPE) is required to reduce the risk of jobsite injuries. Workers need to be informed of all hazardous chemicals that will be used on the project, their potential effects, and emergency and first-aid procedures. A continual safety awareness campaign is needed that is focused on reducing accidents. Frequent safety inspections should be conducted of the jobsite to identify hazards and ensure compliance with job-specific safety rules. Every project meeting should address safety at some level. Many construction firms require their foremen to conduct weekly safety meetings with field craftsmen to maintain a continuous emphasis on hazard removal and safe work practices.

All accidents or near-miss incidents should be thoroughly investigated for lessons learned. The objective is to determine why the accident or incident occurred, and identify procedures or policies to minimize the potential for future occurrence of similar accidents or incidents. A construction firm's safety record has significant impact on its labor cost. Workers' compensation premium rates are adjusted based on the history of claims submitted by the firm's employees.

Case studies

Case studies in this chapter include:

76. Brick detail
77. Subcontractor quality control
78. Carpenter classrooms
79. Bid-design-build
80. Sinking swimming pool
81. Brick mock-up
82. HVAC units
83. Wrong carpet
84. Highway accident

Most of these case studies overlap with other primary topics. Case studies 22, 41, 54, and 60 also involve quality or safety control.

Toolbox quotes

"Active" versus "passive" quality control.

"Active" versus "passive" safety control.

> The motto at a former employer
>
> *We don't have time to do it right, but we always seem to*
> *have time to do it over.*
>
> A quote from a poor contractor
>
> *We can't build it safely, on schedule, within budget, and still*
> *maintain quality – something has to give.*

Case 76: Brick detail

a. This architect is very new to her career. The firm she is working for is quite
    small and specializes in educational facilities. She has been assigned to a public
    university project, which is more than double the value that either she or her
    firm have ever undertaken. This firm is very much a "per plans and specifica-
    tions" architectural firm. For example, she will return submittals 100% reject-
    ed just because five copies were forwarded, not the six that were required in
    the specifications. The general contractor (GC) and the GC's project manager
    who are working on the project have just the opposite background than that
    of the design team. The construction team has mostly private negotiated pro-
    ject experience, and they are very accustomed to large projects. They are the
    largest commercial contractor in the three states that they work. The project
    manager and the architect are not working well together. How would you
    predict this project will conclude? What can the GC's PM do to adjust to this
    type of system?

    _____
    _____
    _____
    _____
    _____

b. On one site visit, this same architect noticed the exposed interior brick lintels
    were not what she had intended. There were not any clear details for this
    work in the bid documents. She did not discuss her concerns with the general
    contractor, but rather returned to her office and discussed the installation with
    her supervisor. Two weeks later, at a weekly construction coordination meet-
    ing, she reported to the owner's representative about this deviation and stated
    she wanted it added to the quality control issues log. The brick mason foreman
    had chosen to repeat another lintel detail that was available in the documents,

although it was for an exterior wall. This soldier course layout he followed was actually more difficult than the detail that the architectural firm desired. The foreman thought he was doing the right thing. What reaction will the general contractor's project manager and superintendent have during the meeting? What will the owner's representative do? Did the subcontractor error? How should the construction team have dealt with this lintel detail in the first place? How should field QC issues be reported by designers and project owners, and to whom and when?

_____

_____

_____

_____

_____

Case 77: Subcontractor quality control

This general contractor was constructing a very sensitive medical tenant improvement (TI) project. The superintendent was very laid back and the project manager was at times combative with the client and the architect. The architect noted two quality control concerns early in the project that warranted more attention. These were the preparation and flashing of roof penetrations and the taping and fireproofing of the vertical drywall surfaces. These conditions occurred around a medical procedure room that required the close attention of all the team members. The superintendent's response to the drywall issues was, "The subcontractor is the expert and I don't tell him what to do." The PM further responded to the roof issues with, "Unless you want to pay me a change order to alter the process, we will stay with our course. If it leaks, the roofing subcontractor will have to return and incur the cost of repairs." Is this a "client-friendly" GC? Are they contractually correct? The architect is trying to implement active QC by bringing up potential problems early. Is this a good practice? Given the GC's response, will the architect continue with early notifications? If this is a "means and methods" issue, can the architect force changes without a change order? Can she issue an American Institute of Architects (AIA) construction change directive (CCD) to force the issue? If the architect's concerns go uncorrected but eventually turn out to be a warranty issue, is the owner in a better position to claim impact against the general? Will the owner or architect choose this GC again?

_____

_____

_____

_____

_____

Case 78: Carpenter classrooms

This project includes new classrooms and training workshops for the carpenters' union apprentices. The general contractor has bid the project lump sum and is approximately 90% through the schedule. Gypsum wallboard (GWB) is being taped and finish materials begin to arrive on the job. On a recent walk-through, the owner and the design teams realize that several items are not exactly as they had anticipated. The gray carpet is actually black. The plywood wainscot in the warehouse area is CDX (construction) grade, not AC (finish) grade. The design-build sanitary waste pipe in the ceiling space between the two floors is plastic and not cast iron. The pipe minimally meets code and will be noisy. The gates on the fence are swinging and not rolling. The interior wood trim is hemlock and not oak. There are many other examples of these types of surprises. The contract requirement for preparing submittals was generic, and although it did list a few items to be submitted, it did not list everything. The architect and project owner acknowledge that there were conflicts in the documents. The contractor has chosen the least expensive materials wherever possible, and is now basing their argument on document inconsistency. The architect is recognizing that there were inconsistencies, but is pleading to the contractor that if there were questions, they were to ask with a request for information (RFI), or a submittal could have been used to verify their intentions. It is now too late to make changes, and the owner will have to live with these conditions. Using these or other specific material examples, explain how project management tools could have been used to prevent these surprises. How could each of the parties improve their performance? Is this a way to achieve repeat negotiated work for any of the parties? If there had been a withholding of funds by the client, who would win the dispute?

Case 79: Bid-design-build

This general contractor has been awarded a two-story speculative office building project on a lump sum bid basis. The mechanical, electrical, and plumbing (MEP) systems, including the heating, ventilation, and air conditioning (HVAC) and fire protection systems, were also bid on a lump sum but design-build basis, with very little criteria information available for them to base their bids on. The subcontractors were responsible for preparing their own documents and having them stamped by a state-licensed engineer and obtain their own permits. The general contractor did not submit these documents to the owner for approval, as it was not specified that they had to do so. The MEP systems subcontractors have routinely received city inspections and approvals for work in place. As the project nears completion, the owner and his architect have just walked the project, and they have discovered that several areas of the design-build MEP subcontractors' work is not up to their expectations. This includes:

- HVAC: The ceiling is being used as a return air plenum. This is much less expensive than utilizing a ducted return air system. It is also noisier, less efficient, dirtier, and requires plenum-rated electrical cabling.
- Plumbing: The bathroom plumbing fixtures appear to be more residential than commercial-grade. They are less expensive, but they do meet code.
- Electrical: The light fixtures that are being installed are 2'×4' prismatic versus more energy-efficient, more expensive, and more attractive deep cell parabolic.
- Fire protection: The fire protection sprinkler heads have not been installed center of ceiling tile, and are not lined up in the large open office areas and hallways.
- MEP controls: The HVAC systems have been ganged under very large control zones such that there are only two thermostats per floor. Even the conference rooms, break rooms, and corner offices are not on separate zones.

The owner's representative is now withholding a current pay request for $300,000, requesting the general contractor correct what he feels are deficiencies. Can he do this? Should the general contractor keep proceeding? Should the GC force the subcontractors to fix the problems? Do subcontractors care about client satisfaction? How will this be resolved? What could the GC have done to keep this from occurring? How would your answer differ if this were a negotiated project versus lump sum bid project?

_____

_____

_____

_____

_____

Case 80: Sinking swimming pool

Maybe you should start this one by developing a timeline. Twenty years ago, this doctor had a swimming pool built on a hillside adjacent to their executive residence by a local reputable general contractor (GC1). The doctor was a wheelchair user. Because of this, he did not inspect the supports under the pool upon completion of the original construction. Ten years later, during a rainstorm, the hillside supporting the swimming pool suffered a slide that may have affected the pool supports, but no one performed an inspection at that time. The pool continued to function as originally constructed.

A geotechnical evaluation was directed five years after the slide as part of a refinancing package. The major support for the cantilever of the pool was three feet below the pool and was not providing any structural support. The study concluded that although the swimming pool may have been constructed properly and had performed well, additional foundation stabilization through a pin-pile technique was recommended. The doctor followed the geotechnical engineer's (E1) recommendation and hired a contractor (GC2) to complete the required work. This second contractor was given all of the recommendations and specifications from the engineering evaluation and was contracted to properly complete the modifications. The contractor invoiced and was paid for work completed. Neither the owner nor the bank nor the engineer inspected the repairs.

Following an earthquake three years later, the doctor contacted a third contractor (GC3) to complete a visual inspection of the swimming pool structure. The contractor inspected the pool and discovered poor workmanship and potentially faulty construction methods used by the second contractor. GC2 apparently failed to positively attach the piles to the supporting beams, relying on gravity and weight instead. GC3 approached an experienced geotechnical/concrete expert to assist with the project. Upon inspection, this second engineer (E2) concurred that the repairs made three years earlier by GC2 were not according to the first soils expert's (E1) recommendation. They now called for new concrete and soils testing to see if the supports could be reattached with a new bracket that could hold everything in place and provide equal support as originally intended.

a. Who should pay for the additional work required to correct the current problem? Is it the original contractor (GC1) who built the pool almost 20 years prior? Is it the second contractor (GC2) who inadequately installed the added supports after the slide? Is it either the first geotechnical engineer (E1) or the bank that didn't inspect the repairs after the slide?

_____

_____

_____

_____

_____

b. Should the homeowner's insurance policy pay for all of the repairs following the recent earthquake? Is the homeowner responsible because he did not adequately inspect the work, and in effect by paying the earlier parties let them off the hook? Is the latest team of contractor (GC3) and engineer (E2) just looking for more work? Do inspecting contractors and designers and financiers have an incentive to find fault with existing conditions? Are they liable if they inspect and do not uncover defects? Is any of the work under warranty? Does this qualify as a "latent defect"?

_____

_____

_____

_____

_____

Case 81: Brick mock-up

The architect for this brick dental facility had specified a requirement for a full-scale mock-up of all the exterior closure elements around a window. This involved four different subcontractors and approximately 10 different types of materials. Are mock-ups an active or passive quality control technique? Are they submittals? How long should a mock-up remain intact on the site? The general contractor on this project attempted to value-engineer the mock-up out of the project. Is this an example of active or passive quality control? Why would they want to delete the mock-up? The brick mason subcontractor used their two best foremen to build the mock-up. Was this a good idea? Who should build mock-ups? The architect prepared a punch list before the mock-up was accepted so that the contractor could continue base contract work. Later, both of the mason's foremen moved onto other projects. The subcontractor had a difficult time repeating the level of quality established in the mock-up, but managed to finish the contract work. The architect and the project owner were pleased with the aesthetic look of the completed brick and window system. Was the designer's intent for this mock-up fulfilled on this project?

_____

_____

_____

_____

_____

Case 82: HVAC units

The general contractor on this project chose not to purchase the heating, ventilation, and air conditioning (HVAC) equipment from the mechanical subcontractor, but rather they purchased it directly from the manufacturer. Why would a GC do this? The shop drawings were submitted and were per specifications. The submittal was approved and returned on time. The equipment was delivered on time, but because the HVAC subcontractor had not yet mobilized on-site, the GC unloaded the mechanical units from the truck and hoisted them to location with the GC's own crane. The subcontractor connected the equipment after placement. The system was completed on schedule, but on a very hot design day, during HVAC balancing, it is discovered that the units are not keeping up with air conditioning requirements. The labels are checked, and it is discovered that they are undersized by one-third from the specified units. The approved submittal does not match up with what has been installed. Who is at fault? How could "active" quality control have prevented this from happening? How is it now resolved contractually and physically?

_____
_____
_____
_____
_____

Case 83: Wrong carpet

This is a small tenant improvement (TI) office project with a contract value of approximately $500,000. The owner is very construction-experienced and routinely has one to two projects under construction at any given time. The architect is a repeat firm. The owner usually likes to negotiate projects, but has recently had two bad experiences with general contractors and decides to bid this project out in hopes of finding a new company they can develop a relationship with. The general contractor who is selected is looking for just this sort of arrangement. They are a very small firm; even this project is at their limit. They operate very informally with little documentation. The project is on schedule and all parties are operating in a negotiated fashion. Samples of the carpet are not submitted as the specifications did not require this process. The carpet arrives two weeks before it is scheduled for installation. It is left rolled up and wrapped. The superintendent does not inspect it. The carpet was installed by the subcontractor over a weekend. The GC's superintendent was not present during this work.

On Monday morning, the owner arrives and is surprised to see that the new carpet is a different color and a different cut than the carpet it adjoins with and was to match. It turns out that the carpet is "as specified," but the architect errored with their specifications. A new replacement carpet is 14 weeks out. The new furniture is coming off the truck and is ready for installation. The owner needs to move new people in this week.

What should the team do? Who pays for any rework? Should contractors submit only what is asked? Why would the design team not request that "everything" be submitted? What happens if a contractor submits on a product for which a submittal was not requested? Will the design team review it? Will the architect's errors and omissions (E&O) insurance pay for removal and replacement and impact costs if this carpet is returned? Does the subcontractor have any liability? Did the project owner find their new general contractor team member?

_____
_____

_____
_____
_____

Case 84: Highway accident

A flagger was killed on a highway construction project when a driver acciden-
tally hit her with his car. Unfortunately, she was not paying attention and was
talking with a parked dump truck driver when the accident occurred. The construction project appeared to be adequately signed and barricaded. The driver was
also injured, and his car is totaled, as he swerved at the last minute to try to avoid
the flagger. The woman was a young mother and her family's insurance company
sues the driver, the contractor, and the state highway department (the client). The
driver's insurance company in turn sues the deceased woman's family, the contractor, and the state highway department. Who can sue whom? We obviously know
who lost in this scenario, but who wins in court? How can highway construction
accidents be mitigated?

_____
_____
_____
_____
_____

12

CHANGE ORDERS

Introduction

The difference between an addendum and a change order is an addendum is issued before the project is bid and a change order is issued after the contract is executed. A change order is an agreement signed by the general contractor (GC), the architect, and the owner modifying some aspect of the scope of work and/or adjusting the contract sum or contract time, or both. Change orders are also known as contract modifications. They may be additive, if they add scope of work, or deductive, if they delete work items. Since they occur on most construction contracts, managing change orders is an important project management (PM) function. Unfortunately, many project disputes begin due to a disagreement on what constitutes a changed condition and/or the cost of that change, as exhibited by the case examples in this chapter and the next on claims.

Change orders originate from a variety of sources and for a variety of reasons. The largest quantity of change orders usually comes from *design errors*, but they are not necessarily due to poor design. Sometimes designers do not have sufficient funds, resources, or time to do a complete and thorough design. Sometimes owners contract with the sub-consultants (such as mechanical or electrical engineers) directly and not through the architect. This was the situation with case studies 1 and 2 presented at the beginning of this book. This may result in multidiscipline documents that are not fully coordinated. And design errors can occur due to the complexity of the project or innocent human error or oversight, as discussed in case study 60. Some design error change orders can be expensive, but often individually they are not costly. Conversely, the most expensive change orders come from owner-directive *scope changes*, such as adding space or changing program or use of a building.

The third type of change order originates from uncovering *unknown site conditions*, which can be the most difficult type of change order for an owner to accept.

These generally result from inadequate site exploration prior to starting design. Hidden or latent conflicts or conditions are common in site work (for example, unknown buried debris) or remodeling (for example, rotten wood structure). When the construction team encounters hidden site conditions that adversely impact construction, the project manager must "promptly" notify the architect and owner, and provide the architect an opportunity to make an inspection.

One type of change order that is avoidable is the lack of coordination of not in contract (NIC), or *owner-furnished equipment* and materials. Owners sometimes believe that they can contract directly with subcontractors and suppliers and save the general contractor's fee. This is common with items such as kitchen equipment, auditorium equipment, furnishings, custom casework, and landscaping. Conflicts may arise and disrupt the project due to the lack of coordination of owner-furnished materials. Owners often will eventually pay much more to resolve these conflicts than they would have paid the general contractors in fees to manage this work.

Change order proposals (COPs) can be initiated by any of the contracting parties, including the project owner, the general contractor, or subcontractors. An owner-initiated proposal request contains a description of the proposed change in the scope of work and requests that the project manager provides an estimate of the cost and time impact on the project. A contractor-initiated proposal describes the proposed change in the scope of work and requests an appropriate adjustment to the contract price and/or time. The best process for a contractor to follow to gain change order proposal approval is to be timely, open-book, and fair with their pricing. An example COP cover sheet is included as Figure 12.1.

The general contractor maintains a change order proposal (COP) log to track all change order proposals, whether generated by the owner or the general contractor and subcontractors. Once the project manager and the owner have negotiated a mutually agreeable adjustment in contract price and time, a formal change order is executed, modifying the contract. If the owner wants to proceed with a change but has not negotiated an appropriate adjustment to the contract price and time, a construction change directive is issued directing the work to be accomplished. The project manager later submits a change order proposal responding to the construction change directive with a proposed adjustment to the contract price and time, but often after costs have been incurred.

Case studies

Case studies included with this chapter are:

85. Demobilized demolition subcontractor
86. Subcontractor change orders
87. Time and material change orders
88. New chief executive officer
89. Dirt change orders

King County Builders, Inc.
4100 SW Hilltop Road
Rainier, WA 98111

CHANGE ORDER PROPOSAL

Project: *Health and Wellness MOB, Project 869*   PM: *Jack Adams*
COP #: *22*   Date: *11/15/2019*

Description of work:
Additional structural steel support beams and acoustical T-bar ceiling modifications to accommodate five owner-provided exam room ceiling lights. Furnish and installation of lights by others.

Referenced documents: *RFI 142, ASI 15, Sk 23, Drawings A441 and S303, Light cut sheets*

COP Estimate Summary: Extended Cost:

1	Direct Labor:	37 hours @ $41/hour	(See att'd)	$1,517
2	Supervision:	0 hours @ $55/hour		$0
3	Labor Burden:	50% of labor:		$759
4	Safety:	2% of labor:		$30
5	Total Labor:			$2,306
6	Direct Materials and Equipment (see attached detailed pricing recaps):			$13,750
7	Small Tools		1% of DL	$15
8	Consumables		1% of DL	$15
9	Total Materials and Equipment:			$13,780
10	Subtotal Direct Work (Items 1 through 9):			$16,086
11	Subcontractors (See attached subcontractor quotes):			$7,200
12	Overhead on Direct Work Items: (included w/fee)			$0
13	Fee on Direct Work Items:		8% of DL & DM	$2,091
14	Fee on Subcontractors:		4% of subcontractors	$288
15	Subtotal Overhead and Fee:			$2,379
16	Subtotal:			$25,665
17	State Excise Tax		0.75% of subtotal	$192
18	Subtotal:			$25,858
19	Liability Insurance:		0.95% of subtotal	$246
20	Total this COP # 22			**$25,911**

This added work has an impact on the overall project schedule, the extent of which cannot be thoroughly analyzed until after the change is incorporated into the contract and the work has been completed.
Please indicate acceptance by signing and returning one copy to our office within five dates of origination.

Approved by:

Dr. Robert Callahan   *11/21/2019*
Physicians and Associates, LLP   Date

FIGURE 12.1   Change order proposal

Most of these case studies overlap with other primary topics, and 20% of the other case studies in this book involve change order proposals to some degree. Specific cases also directly related to change orders include 17, 24–30, 60, 91–93, and others.

> Toolbox quotes
>
> *Change orders exist, so deal with it.*
>
> *The client should pay for everything once, but only once.*
>
> *Sell the change order.*
>
> *Open-book approach to change order management.*

Case 85: Demobilized demolition subcontractor

Your position is the project manager for a general construction firm on a $100 million college laboratory remodel and expansion project. Your demolition subcontractor had a competitive bid of $1 million, just slightly lower than the second bidder. They mobilized quickly and performed the bulk of the original contracted demolition work ahead of schedule. Because of differing site conditions, the architect has requested pricing on numerous change orders affecting the demolition subcontractor's work. The merit of these changes is not in question, but the quantum is. The subcontractor sees this as a "contracting opportunity" and has potentially inflated their pricing on the changes, totaling an additional $1 million. The owner employed a third-party estimating firm to prepare check estimates. The owner offered the subcontractor $500,000 for the changes, which were yet to be performed. The subcontractor did not like the offer and demobilized from the project, having completed all of the base contract work. There appears to be little you can do to motivate them to return. The client has indicated that this is your problem as they made a fair offer for the extra work. What can you do now to remobilize this subcontractor? Should you bring on another construction firm to perform this work? How can you do this contractually? Can you assure that their pricing will be fair? Should you accept the $500,000 change order and hire another subcontractor on a time and material (T&M) basis? Can the owner force you to perform extra work at a value less than what you have requested?

Case 86: Subcontractor change orders

Two subcontractors approached change orders with a medical client differently. This was a complex lump sum project. The documents had errors and the owner also requested scope changes. The mechanical subcontractor did not charge the owner for small discrepancy changes, but successfully collected on the substantial scope changes. The electrical subcontractor charged the owner for every $100 discrepancy. The owner took an immediate dislike to the electrical subcontractor; they burnt their bridge early in the project. The owner was tough on them, even on the clear scope changes. The electrical subcontractor eventually claimed the owner at the end of the project. The mechanical subcontractor made a very fair profit. Was this a good approach from the mechanical subcontractor? Does it always work? Did the electrical subcontractor fail? If so, how did they fail? Where do you draw the line between "give and take"?

_____

_____

_____

_____

_____

Case 87: Time and material change orders

A sole proprietor developer has contracted with a general contractor to build a $40 million condominium complex. The developer has never constructed a project before and does not have relationships with either the contractor or the architect. The architect's services were terminated after receipt of the building permit. The developer acts as his own owner's representative. The contractor was employed through the competitive bid procurement method. There are numerous document discrepancies and differing site conditions. The contractor provides notice to the owner and proceeds with the most likely corrections. The owner is always pushing the contractor to speed up the project under advice from his realtor that the market is hot, and he needs to get the condominiums on the market. The changes are not estimated before the corrections are made, but the contractor tracks their costs on a time and material basis. The contractor does provide the owner with written notices that they are proceeding with each extra work item and will provide actual costs when they are completely realized.

Upon completion of the project, the condominium market has dropped off. The developer is unable to sell 50% of the individual units and ends up unloading the entire project to a large real estate trust at below cost. The remaining condominiums will now be rented as apartments. The contractor has submitted a claim for $1,500,000 for extra work, as described above. All of the extras are legitimate and the pricing is substantiated with T&M backup. The GC has not inflated or falsified any of its costs. Due to the expedited schedule, the project was difficult for the GC. If they collect this claim, they will just break even on their costs. The first developer denies the extras under the basis that: (a) he did not receive proper notification; and (b) he has lost money on the deal as it is, and there is not anything left to share with the contractor. All parties in this case made numerous errors. What were they, and what should they have done to correct this situation before it happened? Does the developer have any claim against his realtor or the architect? Does the developer have a claim against the contractor? Can the GC recover from, or lien, the new owner?

_____

_____

_____

_____

_____

Case 88: New chief executive officer

This is a negotiated team-build assisted living project. It is a complex remodel and addition. The patients are moved from wing to wing as the building is phased over a two-year construction period. The client, architect, general contractor, and major subcontractors all have experience working together. Halfway through the project, the client's chief executive officer (CEO) retires and a new individual is appointed by the board of directors (BOD). The new CEO has lump sum construction experience. Several of his projects have resulted in claims and litigation. He distrusts all parties involved in the construction process, yet he enjoys the battles. He immediately begins to tell the BOD how the previous CEO was not managing the project effectively and that all of the parties are taking advantage of the client. The project is on schedule, quality is adequate, there have not been any safety incidents, but there are numerous change orders outstanding. The CEO convinces the board that he should hire a construction consultant to evaluate the change orders. How should the general contractor's project manager deal with this situation? Should he team up with the other members of the design and construction team against the owner? If the project manager opens his books and works with the estimating consultant, would this help his position? Should the PM make a run on the board? Should he try to win over the new CEO, and if so what approach would you suggest?

_____

_____

_____

_____

_____

Case 89: Dirt change orders

a. A general contractor was the successful bidder on a $15 million lump sum three-building shelled speculative office building complex and has placed a very young and inexperienced team of project manager and superintendent out on the project site. The client is a sole proprietor developer and a very experienced construction professional. The developer advises the contractor's officer in charge (OIC) of his concern of this young GC team. The OIC responds that this is a lump sum project and assignment of personnel is the GC's choice. He also assures the client that he will watch over his team. The project eventually goes astray for several reasons. The contractor loses money on the job and ultimately ends up dismissing both of the on-site staff. There are a significant amount of change orders, both agreed and disputed, and the project finishes three months behind schedule. Can a client in this situation request certain team members or require a change of personnel? Can the client now claim the contractor on the basis of an "I told you so?" What would you have done?

---

b. This same site is against a hillside, has a reputed active salmon stream through it, and has a high water table. In addition, the project ends up with most of the earthwork performed in a record wet and rainy winter. There were three earthwork bids for this project. One subcontractor bids the earthwork at $1 million. The other two are close together at about $300,000. The high bidder is dismissed by the GC's young PM as obviously not needing the work and bidder number two is contracted. It turns out that there were only three soils borings over the entire 10-acre site described in the geotechnical report. The geotechnical firm had advised the developer that it would be better to perform at least 10 borings, but due to cost reasons the developer authorized only three samples. The report indicated that there was "some" water, and the dirt "may not" be suitable for backfill unless kept "dry," and the contractor should anticipate "some" organic materials, but there was not any mention of debris. It turns out that not only is the site very wet, but it is completely a fill site. The earthwork is unsuitable for backfill. The contractors discover substantial debris such as stumps, barrels, garbage, and even a Volkswagen Bug car body. None of the debris is contaminated. Have you ever read a soils report? Is the language specific or vague? What strategies do the geotechnical engineer, client, and even a bidding GC or earthwork subcontractor take with respect to soils reports?

---

c. It turns out that the high-bidding earthwork subcontractor had dumped most of the debris at this site for a prior owner. It was not the subcontractor's workload that caused their high original bid, but rather insider knowledge. What could the general contractor have done to have better anticipated this situation?

Is the young project manager negligent because he did not interview the high bidder? Is either the prior owner or the prior earthwork contractor liable? Is the geotechnical engineer liable in any fashion? There are many potential suspects in this case and in others, as shown in Figure 12.2.

d. The change orders for the earthwork noted above are significant and will eventually amount to $3 million. This is 20% of the original total bid amount from the GC. This of course quickly eats up any of the contingency the developer may have had. Why is earthwork so difficult to estimate? Are earthwork contractors as adept at estimating as are other subcontractors such as glazing or drywall? Do clients carry separate pools of contingency funds, such as some for earthwork and some for mechanical?

e. The inexperienced construction team decides that the best way to keep the original base earthwork contract and the change order earthwork separate is to hire a second earthwork contractor to perform the change order work. Is this customary? The original contracted earthwork firm essentially finishes their work, demobilizes, and is paid 100% of their contract. Later, when the owner claims that some of the change orders should have been assumed by the original subcontractor, there is nothing left of their contract to back charge against. Do either the client or the general contractor now have any recourse against this first subcontractor? The second firm performs the changes utilizing unit prices and time and materials cost accounting. Is this customary on a fast-track project? How should change conditions be performed under a lump sum agreement?

f. The costs and the scope are substantiated for the second earthwork firm, but are they for the general contractor? Does the developer agree to these extra costs? If the developer doesn't agree, can the general contractor refuse payment to the second subcontractor? If this had been a guaranteed maximum price

(GMP) scenario, would the general contractor be allowed to pay for extra work from other sources of savings?

_____

_____

_____

_____

_____

FIGURE 12.2 Potential suspects

13

CLAIMS

Including dispute resolution techniques

Introduction

The complexity of construction contracting occasionally results in project issues that cannot be resolved among the jobsite participants. Such issues from a contractor's perspective typically involve requests for additional money or time for work performed beyond that required by the construction contract. The project manager first submits a change order proposal (COP) for the contract adjustment. If the project owner and general contractor's (GC's) project manager mutually negotiate a COP, a formal contract change order is drafted and a claim will not be necessary. Change order proposals and change order case studies were presented in Chapter 12.

Dispute resolution

If the owner does not agree with the project manager's request, the result may become a contract claim. Procedures for processing claims typically are prescribed in the contract. The normal procedure is for the general contractor's project manager (PM) to formally submit the request for additional compensation and/or time to the owner or the architect along with documentation supporting the general contractor's position. The owner or the architect formally responds to the contractor's request agreeing, agreeing in part, or rejecting the contractor's claim. If the GC's PM does not agree with the response, the claim becomes a contract dispute, and the dispute resolution techniques prescribed by contract are used to settle it. Dispute resolution techniques, from least costly and fastest to the most expensive and longest to resolve, include:

- prevention;
- negotiation;

- mediation;
- dispute resolution boards;
- arbitration; and
- litigation.

It is to the advantage of both the contractor and the owner to resolve the dispute quickly. The farther up the ladder above the project level the dispute reaches, the more likely it is to become adversarial, time-consuming, and expensive.

Negotiation involves both parties sitting down, discussing the issue, and reaching an appropriate resolution. Mediation is an assisted negotiation process in which both parties agree to use a neutral facilitator or mediator. The mediator listens to both parties' positions and attempts to help them reach a consensus regarding resolution. A dispute resolution board (DRB) is a panel of three industry experts convened prior to the start of construction. The project owner, general contractor, and all three board members sign a *three-party agreement*, as reflected in Figure 13.1. The board periodically meets at the project site to review progress on the project and hear any issues in dispute. Either the contractor or the owner can refer an issue to the dispute resolution board for a recommendation. Informal but comprehensive hearings are held and the board makes a recommendation. Board recommendations are not binding on the contractor or owner. A single neutral third party is similar to a DRB, but an even more economical option for the contracting parties.

Arbitration involves both parties formally presenting their case to a neutral third party for a decision. The arbitrator requests written position papers and supporting documentation from both parties to the dispute. Once the documentation has been reviewed, the arbitrator conducts a hearing allowing both sides to present their positions and renders a written decision. Litigation means referring the

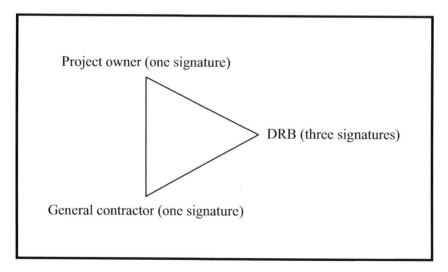

FIGURE 13.1   Three-party agreement

dispute to a court for resolution. This involves hiring legal counsel, preparing necessary documentation, and scheduling an appearance in court. This typically is extremely time-consuming and expensive. Most contractors and owners attempt to resolve their disputes without resorting to litigation, but it might be used as a last resort. Many believe that attorney involvement in any aspect of a construction dispute only drives the parties farther apart and increases cost and time for resolution, as shown in Figure 13.2.

Formal partnering and team-building techniques and others have been adopted by many in the construction industry in an attempt to reduce the number of claims and disputes. The frequent meetings and the issue escalation system are used to resolve issues at the project level in a timely manner.

Case studies

Case studies included with this chapter are:

90. Nice project manager?
91. Written notifications

FIGURE 13.2 Attorneys: If your claim gets to court, there is only one winner, the attorney

92. Plumbing claim
93. School electrician

Most of these case studies overlap with other primary topics. Case studies 23–27, 49, 80, 83, and 87 also involve claims and dispute resolution.

Toolbox quote

Claim is not a four-letter word, but it should be.

Case 90: Nice project manager?

This $80 million lump sum public high school was being built for a school district that had never completed a construction project of this size. The elementary schools and remodel projects undertaken prior were all successful projects. They had never been claimed. The project manager for the school district was experienced with smaller remodel projects, but nothing of this size. The general contractor's (GC's) project manager (PM) was very experienced.

The owner consistently missed deliverable dates (permits, equipment, decisions on finishes), all of which were on the contractor's critical path. When this happened, there was never an issue; rather, the GC's superintendent would simply record it in red on the contract schedule hanging in the jobsite trailer conference room. In addition, when the architect returned submittals late, or lost requests for information (RFIs), or missed meetings, the PM was very understanding and simply documented it in his meeting notes.

When the PM brought on another assistant PM to the job, along with a separate trailer full of empty file cabinets, the client also thought the GC's leadership was being very thorough. There had not been any emotional issues or nasty letters issued throughout most of the project. The documentation from the contractor's side was extensive. Was this really the "nicest project manager" the school district had ever seen? What do you think was going to happen with this project?

Case 91: Written notifications

During the course of construction of a medical office building (MOB), the general contractor's project manager provided numerous verbal directions to the mechanical subcontractor requiring revisions to the schedule and changes to the plans and specifications. Many of these were required to resolve discrepancies and keep the project on schedule, which resulted in schedule compression. The general contractor and the subcontractor disagreed on pricing for some of these extra work items, but ultimately they were all mutually negotiated into the subcontract. The subcontract contained the following clauses, which were pertinent to this case:

i.   Subcontractor is not to perform work for which they have not received written direction.

ii. Subcontractor cannot submit a claim for extra work for which they had failed to provide timely and proper written notification.

iii. Subcontractor cannot submit a claim for damages associated with project delays caused by the owner, owner's agents, or the general contractor.

a.   Upon completion of the project, the subcontractor submitted a claim, through their attorney, for $75,000 worth of damages due to loss of productivity as a result of receiving multiple change orders and resultant schedule compression. The general contractor denied this claim based upon clauses (ii) and (iii) above. The mechanical subcontractor took the case to arbitration, where the arbitrator found in favor of the subcontractor. The arbitrator ruled that the general contractor had, by their actions, waived the "in writing" clause (ii) and the "no damages for delay" clause (iii). The arbitrator determined that the general contractor also invalidated clause (i) when they verbally directed the subcontractor to perform extra work. In effect, the arbitrator has indicated that because the general contractor did not follow the rules, the subcontractor was not required to follow the rules. He indicated the GC was "negligent and acted in bad faith." Is this fair to the subcontractor? Is this fair to the general contractor? What project management principles should both parties have used to prevent this from happening? Is arbitration the fairest way to resolve cases such as this?

---

b.   Assume the role of the general contractor and prepare your case to appeal this ruling to an upper court. Does the contract allow you to appeal? Will this be expensive? Prepare an argument with at least five points utilizing whatever material you can gather, both from within and outside of this book, the classroom, and standard contracts. Feel free to improvise; this GC needs your help!

_____

_____

_____

_____

_____

Case 92: Plumbing claim

During the paper close-out of your project, you receive a very well prepared, very professional claim from your plumbing subcontractor. This subcontractor worked very well with you throughout the project, although they were not the best with their paperwork. The quality of their construction work was acceptable. They received numerous change orders (worth over 20% of their original bid of $600,000) and received approval of the majority of their requests for extra funds and extra time through the course of construction, after mutual negotiation. They signed your change orders. The plumbing project manager you have worked with throughout the project has been relocated and refuses to return your calls. You are now only hearing from their attorney. Assume a standard AIA GC-subcontractor or ConsensusDocs agreement. Their claim is based upon the following five major issues:

  i. Discrepant documentation resulting in extra costs: They received change orders for this work but usually at somewhat reduced value than what they requested. They request an additional $60,000 for this cause.
 ii. The cumulative effect of multiple change orders: This has resulted in extra costs of $100,000 associated with loss of productivity and increase in jobsite general conditions.
iii. Schedule compression resulting in loss of productivity: They request an additional $90,000 for this issue.
 iv. Extended schedule that caused additional field and home office costs of $110,000.
  v. Their accounting records indicate that their actual costs ($1.08 million) overran their final contract value of $720,000 by $360,000, or 50%.

   a. Take the subcontractor's position. Why are you right? Use your project management tools and the contract to prepare your argument.

   _____

   _____

   _____

   _____

   _____

   b. Take the general contractor's position. Why are you right? Use your project management tools and the contract to prepare your argument.

---

c. Be objective. How will this claim be resolved? What should you, as the GC's PM, have done to have prevented this? What changes can you propose to the general contractor's change order and contract modification system to minimize these types of late, after-the-fact claims? What can you do as a specialty contractor project manager to not put your firm in these types of positions? Unfortunately, this case happens too often.

Case 93: School electrician

An electrical subcontractor is filing three simultaneous claims against a public school district. The following facts are available:

- The general contractor's contract with the school district is $10 million on a nine-month remodel project.
- The electrical contractor's original contract value was $1.5 million.
- The claims were filed at the completion of the project with only limited informal notice of any of the issues given prior.
- The electrical subcontractor and the general contractor both bid the job lump sum.
- The contracts that have been executed by all parties were standard AIA documents and were included in the specifications prior to the bid. They were executed without any language changes.
- The general construction contract has $1,000 per day liquidated damages. The general contractor contractually passed that risk onto all of its subcontractors.
- The general contractor and the subcontractors are all bonded.
- The quality of the work and the safety performance were within standards and acceptable.
- The project finished six weeks late, forcing the school into using temporary facilities well into October. The school district has realized real cost damages because of the delays totaling $500,000.
- The reason behind the electrical subcontractor's *first claim* is that the original contract schedule had a reasonable work plan that indicated the electrical

subcontractor's planned manpower peaking for one month at two times the average manpower (eight electricians versus four electricians) over the course of the project. The GC did not manage the project efficiently, used up the float shown in the schedule, and the electrical subcontractor ended up overspending their labor estimate considerably, spending 90% of the actual manpower all during the last two months of the job. The theme of the first claim is *inefficiency due to labor compression.*

- The *second claim* is based upon the fact that the documents were at best 60% complete and were not coordinated between the different design disciplines. The client awarded change orders to the electrical subcontractor through the GC, which increased the subcontractor's original contract value by 50%. The electrical subcontractor and the GC requested more funds throughout the change process, but the client and the design team negotiated them down at every turn. The owner included language in the executed changes, which prohibited the GC from submitting additional claims for the added work. The GC included similar language to the subcontractor. The electrician is claiming *loss of productivity due to the cumulative effect of change orders.*

- The *third claim* occurred halfway through the project during electrical rough-in. The GC discovered asbestos and the project was completely shut down for one month. All parties demobilized and remobilized one month later. A schedule extension was granted, but the contract did not allow for extended general conditions through the use of a *no claim for delay* contract clause. The subcontractor is claiming *loss of labor productivity* due to the fact that many of the key workers were lost during the month shutdown. Despite the "no claim" clause, they are utilizing the Eichleay Formula to request *extended field and home office general conditions,* and *loss of fee potential.*

- The GC has processed all three claims through to the designer, who has forwarded them onto the school district. Neither the GC nor the designer took a strong position either way.

   a. Take the position of the electrical subcontractor. Determine the value of all three claims. Use your estimate and schedule from this or another project as backup. Feel free to "fill in the blanks." Present the claims using information you have learned from your books, your classes, and outside professional experience. Perform outside research to reinforce your claims. You will only get one chance to present your position to the school board at a public hearing.

   _____

   _____

   _____

   _____

   _____

b. Take the position of the school board. Listen to the presentation of the electrical subcontractor. Reject all three claims. Use information you have learned in this book and your classes, as well as substantiating data you research from outside of the classroom. Why didn't your designer or GC come to your rescue? Are they required to? You will only get one chance to present your rejection.

_____
_____
_____
_____
_____

c. Because the school board and the electrical subcontractor could not agree, the contract forced both parties into binding arbitration. Take the position of the arbitration panel. You are to decide in favor of one party or the other on each of the three claims separately. You do not need to find in favor of the same party on all three claims. You are to determine final award values. You are to base your decision on the presentation from each of the parties, information you have learned in your classes, information you research outside of class, and the documents. You are in the business of being hired as an arbitrator and cannot afford to have your positions questioned or overthrown by an upper court. They must be correct and sound.

Arbitration ruling claim 1:

_____

Arbitration ruling claim 2:

_____

Arbitration ruling claim 3:

_____

d. Assume that the contract allowed for alternative dispute resolution (ADR) methods and specified that a dispute resolution board (DRB) be established before contract award. Explain what a DRB is and what benefits it provides over conventional dispute resolution methods. How much would the DRB have cost the parties?

_____
_____
_____
_____
_____

e. Reviewing the presentations of the two parties and the contract documents, explain what "recommended" solution the DRB would make to the two parties. When should this DRB hearing be held? What could all four parties (including GC and designer) have done to keep these issues from occurring or, at a minimum, resolving the issues before formal claims were submitted post project? Could the presence of a DRB from the project's onset have prevented these issues from becoming an after-the-fact claim?

_____

_____

_____

_____

_____

14

ADVANCED TOPICS

Including close-out, BIM, and sustainability

Introduction

This final chapter includes case studies of several advanced topics, including construction close-out, building information modeling (BIM), and sustainability. Some other advanced topics included with earlier chapters were integrated project delivery (IPD), lean construction, and value engineering. This chapter had the most significant changes from previous editions at the request of several readers, and because times are always changing, even in construction, this chapter will likely undergo significant "reconstruction" in future editions as well.

Close-out

Close-out of a construction project is the holistic and complicated process of completing *all* of the construction tasks and all documentation required to close out the contract and consider the project complete. As physical construction on a project nears completion, or before, the project manager develops a project close-out plan to manage the numerous activities involved in closing out the project. Just as the project start-up activities, described by case studies in Chapter 7, are essential when initiating work on a project, good project close-out procedures are essential for timely completion of all contractual requirements and receipt of final payment. Not only is efficient project close-out good for the general contractor, but it is also good for the project owner. Owners are often disappointed by contractors that performed well throughout construction but dropped the baton during close-out, as described in case studies in this chapter. The project manager and superintendent want to close out a project quickly and move onto another project. Minimizing the duration of close-out activities generally maximizes the profit on a project, as it limits project overhead costs and facilitates early receipt of final payment and any retention. Expedient close-out and turnover procedures minimizes the contractor's interference with the owner's move-in and start-up activities, and therefore results in a satisfied customer.

When the project is nearing completion, the general contractor (GC) asks the owner and architect to conduct a pre-final inspection. Any deficiencies noted during this inspection are placed on the punch list for future reinspection. All deficiencies on the punch list must be corrected before the contract can be closed out. A significant project milestone is achieving substantial completion, which indicates that the project can be used for its intended purpose. The architect decides when the project is substantially complete and issues a certificate of substantial completion. Even though the project is substantially complete, the owner cannot move in until a certificate of occupancy (C of O) is issued by the local permitting authority.

The project manager (PM) is responsible for the financial and contractual close-out of the project. This involves issuing final change orders to subcontractors and major suppliers and securing final and unconditional lien releases from them. As-built drawings, operation and maintenance manuals (O&Ms), warranties, and test reports must be submitted. The project manager should develop a close-out log early in the project to manage the timely submission of all close-out documents. The objective is to close out all project activities expeditiously so the contractor can receive the final payment and release of retention.

Most construction contracts contain a warranty provision that requires the contractor to repair any defective work or replace any defective equipment identified by the owner within one year after substantial completion. Similar provisions should be included in all subcontracts to cover work performed by subcontractors. Subcontractor warranty documentation is collected during project close-out. Warranty claims need to be investigated and resolved. Generally, the project owner submits any claims to the general contractor, who either notifies the appropriate subcontractors or sends an appropriate craftsperson to investigate. Warranty response is an important aspect of customer service. Poor warranty support, just like a slow close-out phase, may jeopardize the contractor's ability to obtain future projects from the project owner.

Building information modeling

Building information modeling, or models (BIM), has become commonplace on many larger and more complex construction projects. BIM has been very beneficial to design teams to present their early concepts to their clients. Contractors have also used BIM successfully for presentations, estimating, and scheduling. Now design and construction teams are using BIM and other electronic tools through the course of construction to process requests for information (RFIs), submittals, change orders, and other traditionally paper construction communication tools. But BIM has its costs as well, and is not accepted by all in the built environment industries, as reflected in some of these case studies.

Sustainability

The objective of sustainability in the built environment is to plan and execute a construction project in such a manner as to minimize the adverse impact of the construction process on the environment. While much of the sustainable aspects of a completed project are a result of design decisions, other aspects are due to the manner in which the construction is performed. For many years, sustainability was synonymous with Leadership in Energy and Environmental Design (LEED). "By definition, sustainability is the ability of the current generation to meet its own needs without compromising the ability of future generations to meet their needs" (LEED G.A.). There are many other terms or buzzwords that characterize sustainability: *going green*, *the green movement*, and *zero-carbon footprint* are just a few examples. There are also many other sets of sustainability guidelines and certificates, including Green Globes, HERS, and others. Figure 14.1 is a LEED gold plaque proudly displayed in a recently completed university classroom and laboratory building.

When the sustainability movement began, contractors viewed it as burdensome, but now it just makes sense not only in construction, but in everyday life. Originally, it did cost more, but today new materials and improved building methods make it not only reasonable to be green, but it is actually more economical. Design professionals and contractors will incorporate green products and practices in new buildings, even if it is not an owner-generated requirement, as exemplified by one of these case studies. Many municipalities are also adopting green requirements in their building codes, so in the future it will not even be an option.

Case studies

Case studies in this chapter include:

94. Close-out documentation
95. Who closes?
96. No contractual ties

FIGURE 14.1 LEED gold

LEED® and its related logo, is a trademark owned by the U.S. Green Building Council® and is used with permission (usgbc.org/LEED)

97. BIM University
98. Specified BIM
99. Team green
100. Green employees
101. Sustainable development

Most of these case studies overlap with other primary topics. Cases that also involve advanced topics such as IPD and lean and safety include 4, 22, 54, 60, 68, 69, and 80.

Toolbox quotes

It's not how you start the race, it's how you finish that counts.

Make your fortune on the next job, not on this one.

Case 94: Close-out documentation

A general contractor has successfully completed building a shipping and receiving facility. This project was initially procured as a competitive lump sum bid project. This was a fairly simple $10 million, 100,000 square foot tilt-up concrete facility. Total construction time was seven months. The owner, contractor, and architect have all worked well together throughout the project. All payment requests have been approved and paid on time. All change orders have been negotiated and incorporated into the contract. This owner apparently will have other similar projects coming up throughout the country and is interested in hiring the general contractor on a negotiated basis to provide all of their construction needs. The contractor has combined the last progress payment request for $100,000 with a request for retention release of $250,000 for a total $350,000 final payment request. The general contractor moved their project manager off the project during the last month of the schedule and has turned over the close-out of the project to a new project engineer (PE) fresh out of college. The project engineer has had a difficult time getting the close-out documents together, especially the operation and maintenance manuals. The subcontractors are not responding with all of the required information. The architect has twice rejected the O&Ms. Two months have now passed since completing the punch list work and receiving the C of O from the city. The owner is now refusing the requested payment for $350,000. The architect has suggested that the general contractor separate out the last month's $100,000 that is due and request that under a separate payment invoice. The project engineer is being stubborn and is refusing. What mistakes has the general contractor made? What can they do now to remedy the situation? Can the owner contractually hold the full $350,000, and if so for how long? When will the general contractor's (and the subcontractors') lien rights expire? Is there any chance these parties can (or should) enter into a long-term national agreement?

_____

_____

_____

_____

_____

Case 95: Who closes?

This typical school district was feeling the common pains of late projects and uncompleted punch lists that drag on well into the school year. The third-party consultant who served as the owner's representative did not have an incentive to stay with the project after the certificate of occupancy was received. The general contractor was simply looking for release of their retention. The school district's facility managers and maintenance crew did not have any construction involvement and did not have any contractual relations with any of the construction team. Most of the designer's fee is received prior to the permit stage and very little is left for the close-out process. Who should be responsible for proper close-out of a construction project? Do all of the team members have a built-in incentive for a speedy and sometimes poorly performed close-out phase? What about involving the school principals – don't they have the long-term incentive for a properly completed project? Should the end user on any project be involved from design through construction and marshal a properly closed-out project? But are they qualified? Are you, as a recent college graduate PE, qualified to take the point on close-out?

_____

_____

_____

_____

_____

Case 96: No contractual ties

An out-of-town investment firm purchases a speculative office building complex while the project is under construction. The buyer's contract is with the seller. The buyer does not have any contractual ties with either the contractor or the architect. The plans and specifications are part of the purchase-sale agreement. The escrow close occurs upon receipt of certificate of occupancy and completion of all punch list work. The punch list was developed by the seller. The buyer did not participate in development or sign-off of the punch list. Two years after the close, there are numerous problems that show up associated with roof drains, storm drains, and the general slopes of site pavement and sidewalks. The contractor's one-year warranty has expired. Upon investigation, the buyer is notified by the original civil engineer that the project was not built according to the documents in several instances. The engineer did not participate in development of the punch list, as this was not in their contract. Does the buyer have a case against any of these parties? How should purchasers of in-process or recently completed construction projects position themselves to protect against latent defects?

_____

_____

_____

_____

_____

Case 97: BIM University

A private university teamed with a national architect and a major local GC to design and build a 100,000 square foot high-technology laboratory and classroom building. All parties committed to achieve as many of the leading-edge current design and construction management topics as possible, including lean construction techniques, certified LEED gold, just-in-time (JIT) deliveries, locally sourced building materials, and others. Because the students on this campus are so "green" and "techy," the team also committed to design and build the $60 million project paperless. The architect and GC convinced the owner's representative that because of all of these cutting-edge approaches, there was no way to fix a lump sum or guaranteed maximum price (GMP) construction cost estimate up front. The GC therefore received a cost-plus percentage fee (CPPF) contract with a 4% fee. There were not any physical copies of drawings, specifications, RFIs, submittals, meeting notes, job diaries, or other typical construction management (CM) tools utilized throughout the course of construction.

Building information modeling was used by all team members and each company had a full-time BIM operator whose training was paid as a job cost. The BIM operators had their own weekly coordination meeting outside of the other standard owner-architect-contractor (OAC), RFI, change order proposal (COP), submittal, and other weekly jobsite meetings. The project was completed safely, on time, and with adequate quality. The final cost was approximately $750 per square foot. Because of electronic tools, such as BIM, the GC reported upon completion the project cost 10% less than if it had been built utilizing conventional paper CM tools. The wages for the BIM operators for the general contractor, architect, project owner, and mechanical, electrical, and plumbing (MEP) subcontractors amounted to $1.5 million over the course of this two-year project, not including computers and software and office space.

During close-out, the university conducted a standard open-book audit of the contractor's books. The auditor was unaware that the project did not have some price guarantee and was at a loss on how to report back to the university's board of regents that the actual costs could be substantiated. Apparently, the owner's representative, although experienced and authorized to award and manage the contract, did not run the CPPF idea past his superiors.

Can the GC's cost-effective claim be verified? What are the cost differences between a conventional and paperless construction project? Who ultimately paid for all of the BIM coordination on this project? How much did BIM save from the bottom line? How much did it cost? Can a true cost comparison of any two projects ever be conducted? Aren't there always differences?

_____

_____

_____

_____

_____

Case 98: Specified BIM

This architectural design partnership convinced a school district to pay them an additional $1 million in design fees to utilize building information modeling for all architecture and engineering documentation on this $100 million public high school bid project. The architect provided the school district with several academic papers indicating BIM would improve design coordination and reduce the quantity of change orders. Many of the school district's prior construction projects had finished in claims and lawsuits due to change orders, so they were willing to try this new approach to design. The architect included a clause in the special conditions of the contract requiring bidding GCs to include $50,000 in their bid for "BIM support during the construction process." The successful low-bidding GC indicated it had included this $50,000, but because it was a lump sum bid project, the GC's estimate and cost accounting records were closed-book.

It turns out that the project had many scope additions and design discrepancy errors. The GC and its subcontractors submitted $3 million and $7 million, respectively, for these changes. The GC and the subcontractors clearly used BIM efficiently when preparing RFIs and change order proposals as supporting documentation. The GC and its MEP and civil subcontractors forwarded their last COP to the architect for $100,000 to submit as-built drawings electronically. They had planned to redline a set of prints the old-fashioned way, arguing "BIM during construction" was not BIM during close-out. For an additional $25,000, they would complete their O&Ms electronically as well, otherwise the school district would receive five three-ring binders of paper. The architect and owner have rejected these two COPS and are holding all retention, which amounted to $5 million, until electronic close-out documents are provided from the construction team.

Needless to say, the school district was very displeased with all of the parties. During the course of construction, the architect had reduced the proposed values of the scope COPs by $1 million and had rejected $3 million worth of the discrepancy change requests – arguing that if the GC had used BIM as intended, and managed the project properly, these costs would not have occurred. The GC filed a $4 million claim for the COPs, plus the $5 million retention due, plus attorney fees and interest. The school district also filed a simultaneous claim against the architect's errors and omissions (E&O) insurance. It turns out the architect did not require its sub-consultants and engineers to carry E&O as they were required to by the project owner.

Take a position, school district, architect, or the GC, and argue your case. Why are you correct? Which party errored? Do subcontractors or design sub-consultants have any culpability? Can the architect prove their assertion that if BIM is used properly by contractors, the quantity and cost of COPs will be reduced? How will this be resolved? Rolling the clock back two years, what should each of the parties have done to have prevented this?

_____

_____

_____

_____

_____

Case 99: Team green

Many project teams are building in a sustainable fashion today. Even if the entire team is not dedicated to achieving a sustainability certificate, such as LEED, Green Globes, HERS, and others, many of the individual team members are designing and building in a "green" fashion. Today, it is often more economical to build sustainably, and many permitting agencies are also incorporating sustainable materials and methods into their building codes. But there can be issues in achieving sustainability goals, as described in the cases below. We will use LEED as the matrix in these examples.

a.   This project owner wanted a LEED gold certificate for his new building. The architect was to specify accordingly, but it turns out there were not enough points allowed and the project only achieved LEED silver. The general contractor earned all of the points it was responsible for. Is there any action the owner has against the architect now?

_____

_____

_____

_____

_____

b. This project owner was not originally interested in paying additional design and construction and consultant costs for LEED, let alone buying the LEED certificate itself. After move-in, the owner's employees petitioned her to become "green" and certify the new building as sustainable. Is there anything that can be done now? Isn't LEED-certified (less than silver or gold or platinum) fairly easy to attain?

_____

_____

_____

_____

_____

c. This project was designed with enough points in it, but the GC did not track its material sources or keep recycling records adequately. The project achieved LEED-certified in lieu of silver, but the client is not pleased. Can the project owner take any financial or legal action against the GC?

_____

_____

_____

_____

_____

d. Even if a project is designed and built green, it will not remain so if the project owner does not maintain and operate it accordingly. What are some of the ongoing actions a project owner must do to maintain their LEED gold certificate? Do they need ongoing services from a sustainability consultant? Is there a cost to maintain a sustainable certificate? Is there a payback?

_____

_____

_____

_____

_____

Case 100: Green employees

Your construction company has several young project engineers, project managers, and superintendents. Over half of them are certified as Leadership in Energy and Environmental Design Green Associates or Applied Professionals (LEED GA or LEED AP). They want to build "green" even if it is not a contract requirement of the project owner, architect, or city. The company feels this is morally the environmental thing to do. Prepare a list of at least 10 methods and materials your company might incorporate on every project to build green as a course of standard business practice. Does this methodology save money? Can you achieve LEED, or other green rating system project certifications, even without the support of the architect or project owner? Would you do this?

_____

_____

_____

_____

_____

Case 101: Sustainable development

It was this developer's goal to build the most sustainable office building ever, and have it also be a profitable real estate investment project. His goal was to achieve the *living building challenge* of a *zero-carbon footprint*. The design and construction team easily accomplished LEED platinum certification. The five-story spec office building over two stories of underground parking garage cost $1,200 per square foot. It won several local and national architectural and sustainability awards. The project was a huge success, and the developer felt he had set the standard for others to follow and was proud of the investment and legacy he had established for his children.

Three years later, the building is still 75% vacant. The rent the developer needs to cover his financial pro forma is so high that although potential tenants appreciate his sustainability accomplishments, they cannot afford the price tag. The potential tenants also have strict sustainability maintenance rules and practices they must adhere to. Was this a successful real estate development project? Starting all over, how could all of the developer's goals be accomplished? What would you recommend he do now?

_____

_____

_____

_____

_____

APPENDIX A

Abbreviations

The use of abbreviations and acronyms is standard in the construction industry and many are used throughout this book; all of them are listed here along with a few others.

ABC	activity-based costing
AC	finish-grade plywood
ADR	alternative dispute resolution, including *DRB*s and mediation
AGC	Associated General Contractors of America
AIA	American Institute of Architects
AKA	also known as
AP	Applied Professional (*LEED* certification)
Arch	architect
BC	back charge
BIM	building information models or modeling
Board	board of directors or dispute resolution board, *DRB*
BOD	board of directors, also *board*
CCA	construction change authorization
CCD	construction change directive
CDs	construction documents or contract documents
CDX	construction-grade plywood
CEO	chief executive officer
CFO	chief financial officer
CM	construction manager or construction management (person, company, or delivery method)
CM/GC	construction manager/general contractor (delivery method), also known as *GC/CM* or cm at risk
C of O	certificate of occupancy
CO	change order

C-O	close-out
Contractor	general contractor or subcontractor
COO	chief operating officer
COP	change order proposal
CP	change proposal
CPFF	cost plus fixed fee
CPPF	cost-plus percentage fee (similar to *T&M*)
CSI	Construction Specifications Institute
DB or D-B	design-build (delivery method)
DBE	disadvantaged-owned business enterprise
DBOM	design-build-operate-maintain (delivery method)
DDs	design-development documents
Demo	demolition or demolition contractor
DeMob	demobilization
DRB	Dispute Resolution Board
E	engineer
E&O	errors and omissions (insurance)
FE	field engineer
FO	field order
FQ	field question, also known as request for information, *RFI*
FWO	field work order
GA	Green Associate (*LEED* certification)
GC(s)	general contractor or general contractors
GCs	general conditions
GC/CM	general contractor/construction manager (procurement and contracting method), also *CM/GC*
General	general contractor, also *GC* or contractor
GMP	guaranteed maximum price (contract or estimate), also *GMC*
GWB	gypsum wallboard (subcontractor or material), also known as drywall or sheetrock
HERS	Home Energy Rating System
HVAC	heating, ventilation, and air conditioning (contractor or subcontractor or equipment)
IPD	integrated project delivery
ITB	instructions to bidders or invitation to bid
JIT	just-in-time (material delivery)
JV	joint venture
K	1,000 ($40k = $40,000)
LDs	liquidated damages
LEED	Leadership in Energy and Environmental Design
LR	lien release
LS	lump sum (agreement, bid, estimate, contract, or process)
M	million (dollars), also *mil*
M&E	mechanical and electrical (subcontractors or scope), also *MEP*

MACC	maximum allowable construction cost, similar to *GMP* and *GMC*
MBE	minority-owned business enterprise
MEP	mechanical, electrical, and plumbing (subcontractors or systems)
Mil	million (dollars), also *M*
MOB	medical office building
Mob	mobilization
MXD	mixed-use development
NA	not applicable or not available
NIC	not in contract, also not included
NTP	notice to proceed
O&Ms	operation and maintenance manuals
OAC	owner-architect-contractor (meeting)
OH&P	overhead and profit
OIC	officer in charge
OSHA	Occupational Safety and Health Administration
PE	pay estimate, also pay request, *PR*
PE	project engineer
PL	punch list, also punch
PM	project manager or project management
PO	purchase order (contract agreement for a material supplier)
PPE	personal protective equipment (safety)
PPP	public–private partnership (delivery method)
PR	pay request or public relations, also pay estimate, *PE*
Pre-Con	preconstruction (contract, agreement, process, fee, or phase)
QC	quality control
QTO	quantity take-off
Rebar	reinforcement steel (in concrete)
Rep	owner's representative
RFI	request for information (also known as field question, *FQ*)
RFP	request for proposal
RFQ	request for quotation or request for qualifications
ROM	rough order of magnitude (estimate)
SBE	small business-owned enterprise
SCHD	schedule
SDs	schematic design (documents)
SF	square foot
SK	sketch (numbered supplemental design drawing)
SOV	schedule of values (pay estimate backup or summary estimate)
Spec	specification or speculation (as in real estate speculation)
SPM	senior project manager
Sub(s)	subcontractor or subcontractors or subcontract
Super	superintendent, also supt
T&M	time and materials (cost tracking or contract method), also *CPPF*
TI	tenant improvements

TQM	total quality management
UP	unit price (cost estimate or contract)
U.S.	United States
VBE	veteran-owned business enterprise
VE	value engineering
VP	vice president
WBE	woman- or women-owned business enterprise
WBS	work breakdown structure

APPENDIX B

Glossary

Active quality management program A process that anticipates and prevents quality control problems rather than just responding to and correcting deficiencies.

Activity duration The estimated length of time required to complete an activity.

Addenda Additions to or changes in bid documents issued prior to bid and contract award.

Agency construction management delivery method A delivery method in which the owner has three contracts: one with the architect, one with the general contractor, and one with the construction manager. The construction manager acts as the owner's agent but has no contractual authority over the architect or the general contractor.

Agreement A document that sets forth the provisions, responsibilities, and the obligations of parties to a contract. Standard forms of agreement are available from professional organizations.

Allowance An amount stated in the contract for inclusion in the contract sum to cover the cost of prescribed items, the full description of which is not known at the time of bidding. The actual cost of such items is determined by the contractor at the time of selection by the architect or owner, and the total contract amount is adjusted accordingly.

Alternates Selected items of work for which bidders are required to provide prices.

Alternative dispute resolution A method of resolving disagreements other than by litigation.

Amendments See *addenda*.

American Institute of Architects A national association that promotes the practice of architecture and publishes many standard contract forms used in the construction industry.

Application for payment   See *payment request*.

Arbitration   Method of dispute resolution in which an arbitrator or a panel of arbitrators evaluates the arguments of the respective parties and renders a decision.

Arrow-diagramming method   Scheduling technique that uses arrows to depict activities and nodes to depict events or dates.

As-built drawings   Contractor-corrected construction drawings depicting actual dimensions, elevations, and conditions of in-place constructed work.

As-built estimate   Assessment in which actual costs incurred are applied to the quantities installed to develop actual unit prices and productivity rates.

As-built schedule   Marked up, detailed schedule depicting actual start and completion dates, durations, deliveries, and restraint activities.

Associated General Contractors of America   A national trade association primarily made up of construction firms and construction industry professionals. It publishes many standard contract forms used in the construction industry.

Back charge   General contractor charge against a subcontractor for work the general contractor performed on behalf of the subcontractor.

Bar chart schedule   Time-dependent schedule system without nodes that may or may not include restraint lines.

Bid bond   A surety instrument that guarantees that the contractor, if awarded the contract, will enter into a binding contract for the price bid and provide all required bonds. A commonly used form is AIA document A310.

Bid security   Money placed in escrow, a cashier's check, or bid bond offered as assurance to an owner that the bid is valid and that the bidder will enter into a contract for that price.

Bid shopping   Unethical general contractor activity of sharing subcontractor bid values with the subcontractor's competitors in order to drive down prices.

Bridging delivery method   A hybrid of the traditional and the design-build delivery methods; the owner contracts with a design firm for the preparation of partial design documents, and then selects a design-build firm to complete the design and construct the project.

Builder's risk insurance   Protects the contractor in the event that the project is damaged or destroyed while under construction.

Building information models or modeling   Computer design software involving multidiscipline three-dimensional overlays improving constructability and reducing change orders.

Build-operate-transfer delivery method   A delivery method in which a single contractor is responsible for financing the design and construction of a project and is paid an annual fee to operate the completed project for a period of time, such as 30 years.

Buyout   The process of awarding subcontracts and issuing purchase orders for materials and equipment.

Buyout log   A project management document that is used for planning and tracking the buyout process.

Cash flow curve A plot of the estimated value of work to be completed each month during the construction of a project.

Cash-loaded schedule A schedule in which the value of each activity is distributed across the activity, and monthly costs are summed to produce a cash flow curve.

Certificate of insurance A document issued by an authorized representative of an insurance company stating the types, amounts, and effective dates of insurance for a designated insured.

Certificate of occupancy A certificate issued by the city or municipality indicating that the completed project has been inspected and meets all code requirements.

Certificate of substantial completion A certificate signed by the owner, architect, and contractor indicating the date that substantial completion was achieved.

Change order Modifications to contract documents made after contract award that incorporate changes in scope and adjustments in contract price and time. A commonly used form is AIA document G701.

Change order proposal A request for a change order submitted to the owner by the contractor, or a proposed change sent to the contractor by the owner requesting pricing data.

Change order proposal log A log listing all change order proposals indicating dates of initiation, approval, and incorporation as final change orders.

Claim An unresolved request for a change order.

Close-out The process of completing all construction and paperwork required to complete the project and close-out the contract.

Close-out log A list of all close-out tasks that is used to manage project close-out.

Commissioning A process of assuring that all equipment is working properly and that operators are trained in equipment use.

Conceptual cost estimate Cost estimates developed using incomplete project documentation.

Conditional lien release A lien release that indicates that the issuer gives up his or her lien rights on the condition that the money requested is paid.

ConsensusDocs Family of contract documents that takes the place of the AGC contract documents.

Constructability analysis An evaluation of preferred and alternative materials and construction methods.

Construction change directive A directive issued by the owner to the contractor to proceed with the described change order.

Construction manager at risk delivery method A delivery method in which the owner has two contracts: one with the architect and one with the construction manager/general contractor. The general contractor is usually hired early in the design process to perform preconstruction services. Once the design is completed, the construction manager/general contractor constructs the project.

Construction manager/general contractor delivery method See *construction manager at risk delivery method*.

Construction Specifications Institute The professional organization that developed the original 16-division MasterFormat that is used to organize the technical specifications; today's CSI includes 49 divisions.

Contract A legally enforceable agreement between two parties.

Contract time The period of time allotted in the contract documents for the contractor to achieve substantial completion; also known as project time.

Coordination drawings Multidiscipline design drawings that include overlays of mechanical, electrical, and plumbing systems with the goal of improving constructability and reducing change orders.

Corrected estimate Estimate that is adjusted based on buyout costs.

Cost codes Codes established in the firm's accounting system that are used for recording specific types of costs.

Cost estimating Process of preparing the best educated anticipated cost of a project given the parameters available.

Cost-plus contract A contract in which the contractor is reimbursed for stipulated direct and indirect costs associated with the construction of a project and is paid a fee to cover profit and company overhead.

Cost-plus contract with guaranteed maximum price A cost-plus contract in which the contractor agrees to bear any construction costs that exceed the guaranteed maximum price unless the project scope of work is increased.

Cost-plus-fixed-fee contract A cost-plus contract in which the contractor is guaranteed a fixed fee irrespective of the actual construction costs.

Cost-plus-percentage-fee contract A cost-plus contract in which the contractor's fee is a percentage of the actual construction costs, also known as a *time and materials contract*.

Cost-reimbursable contract A contract in which the contractor is reimbursed stipulated direct and indirect costs associated with the construction of a project. The contractor may or may not receive an additional fee to cover profit and company overhead.

Craftspeople Non-managerial field labor force who construct the work, such as carpenters and electricians.

Critical path The sequence of activities on a network schedule that determines the overall project duration.

Daily job diary A daily report prepared by the superintendent that documents important daily events, including weather, visitors, work activities, deliveries, and any problems; also known as daily journal or daily report.

Davis–Bacon wage rates Prevailing wage rates determined by the U.S. Department of Labor that must be met or exceeded by contractors and subcontractors on federally funded construction projects.

Design-build delivery method A delivery method in which the owner hires a single contractor who designs and constructs the project.

Design-build-operate delivery method A delivery method in which the contractor designs the project, constructs it, and operates it for a period of time, for example 20 years. Sometimes known as design–build–operate–maintain.

Detailed cost estimate Extensive estimate based on definitive design documents. Includes separate labor, material, equipment, and subcontractor quantities. Unit prices are applied to material quantity take-offs for every item of work.

Direct construction costs Labor, material, equipment, and subcontractor costs for the contractor, exclusive of any markups.

Dispute A contract claim between the owner and the general contractor that has not been resolved.

Dispute Resolution Board A panel of experts selected for a project to make recommendations regarding resolution of disputes brought before it.

Earned value A technique for determining the estimated or budgeted value of the work completed to date and comparing it with the actual cost of the work completed. Used to determine the cost and schedule status of an activity or the entire project.

Eichleay Formula A complicated method of potentially recovering home office overhead and lost fee potential, usually associated with claims involving time extensions.

Eighty-twenty rule On most projects, about 80% of the costs or schedule durations are included in 20% of the work items; also known as Pareto's 80-20 rule.

Errors and omissions insurance Protects design professionals from financial loss resulting from claims for damages sustained by others as a result of negligence, errors, or omissions in the performance of professional services; also known as professional liability insurance.

Estimate schedule Management document used to plan and forecast the activities and durations associated with preparing the cost estimate. Not a construction schedule.

Exhibits Important documents that are attached to a contract, such as a summary cost estimate, schedule, and document list.

Expediting Process of monitoring and actively ensuring vendor's compliance with the purchase order requirements.

Expediting log A spreadsheet used to track material delivery requirements and commitments.

Experience modification rating A factor unique to a construction firm that reflects the company's past claims history. This factor is used to increase or decrease the company's workers' compensation insurance premium rates.

Fast-track construction Overlapping design and construction activities so that some are performed in parallel rather than in series; allows construction to begin while the design is being completed; also known as phased construction.

Fee Contractor's income after direct project and jobsite general conditions are subtracted. Generally includes home office *overhead* costs and *profit*.

Field engineer Similar to the project engineer, except with less experience and responsibilities; may assist the superintendent with technical office functions.

Field question See *request for information.*

Filing system Organized system for storage and retrieval of project documents.

Final completion The stage of construction when all work required by the contract has been completed.

Final inspection Final review of the project by owner and architect to determine whether final completion has been achieved.

Final lien release A lien release issued by the contractor to the owner or by a subcontractor to the general contractor at the completion of a project indicating that all payments have been made and that no liens will be placed on the completed project.

Float The flexibility available to schedule activities not on the critical path without delaying the overall completion of the project.

Foreman Direct supervisor of craft labor on a project; works for both general contractors and subcontractors and usually reports to a superintendent.

Front-loading A tactic used by a contractor to place an artificially high value on early activities in the schedule of values to improve cash flows.

General conditions A part of the construction contract that contains a set of operating procedures that the owner typically uses on all projects. They describe the relationship between the owner and the contractor, the authority of the owner's representatives or agents, and the terms of the contract. The general conditions contained in AIA document A201 are used by many owners.

General contractor The party to a construction contract who agrees to construct the project in accordance with the contract documents.

General liability insurance Protects the contractor against claims from a third party for bodily injury or property damage.

Geotechnical report A report prepared by a geotechnical engineering firm that includes the results of soil borings or test pits and recommends systems and procedures for foundations, roads, and excavation work; also known as a *soils report.*

Guaranteed maximum price contract A type of cost-plus contract in which the contractor agrees to construct the project at or below a specified cost.

Indirect construction costs Expenses indirectly incurred and not directly related to a specific project or construction activity, such as home office *overhead.*

Invitation to bid A portion of the bidding documents soliciting bids for a project.

Job description A description of the major tasks and duties to be performed by the person occupying a certain position.

Job hazard analysis The process of identifying all hazards associated with a construction operation and selecting measures for eliminating, reducing, or responding to the hazards.

Jobsite general conditions costs Field indirect costs that cannot be tied to an item of work, but which are project-specific, and in the case of cost-reimbursable contracts are considered part of the cost of the work.

Job specifications The knowledge, abilities, and skills a person must possess to be able to perform the tasks and duties required.

Joint venture A contractual collaboration of two or more parties to undertake a project.

Just-in-time delivery of materials A material management philosophy in which supplies are delivered to the jobsite just in time to support construction activities. This minimizes the amount of space needed for on-site storage of materials.

Labor and material payment bond A surety instrument that guarantees that the contractor (or subcontractor) will make payments to his or her craftspeople, subcontractors, and suppliers. A commonly used form is AIA document A312.

Lean construction Process to improve costs incorporating efficient methods during both design and construction; includes *value engineering* and *pull planning*.

LEED Measure of sustainability administered by the U.S. Green Building Council, usually associated with receipt of a LEED certificate.

Letter of intent A letter, in lieu of a contract, notifying the contractor that the owner intends to enter into a contract pending resolution of some restraining factors, such as permits or financing.

Lien A legal encumbrance against real or financial property for work, material, or services rendered to add value to that property.

Lien release A document signed by a subcontractor or the general contractor releasing its rights to place a lien on the project.

Life cycle cost The sum of all acquisition, operation, maintenance, use, and disposal costs for a product over its useful life.

Liquidated damages An amount specified in the contract that is owed by the contractor to the owner as compensation for damaged incurred as a result of the contractor's failure to complete the project by the date specified in the contract.

Litigation A court process for resolving disputes.

Long-form purchase order A contract for the acquisition of materials that is used by the project manager or the construction firm's purchasing department to procure major materials for a project.

Lump sum contract A contract that provides a specific price for a defined scope of work; also known as fixed-price or stipulated-sum contract.

Markup Percentage added to the direct cost of the work to cover such items as overhead, fee, taxes, and insurance.

MasterFormat A 16-division numerical system of organization developed by the Construction Specifications Institute that is used to organize contract specifications and cost estimates.

Materialman's notice A notice sent to the owner as notice that the supplier will be delivering materials to the project.

Material safety data sheets Short technical reports that identify all known hazards associated with particular materials and provide procedures for using, handling, and storing the materials safely; also known as safety data sheets.

Material supplier   Vendor who provides materials but no on-site craft labor.

Mediation   A method of resolving disputes in which a neutral mediator is used to facilitate negotiations between the parties to the dispute.

Meeting agenda   A sequential listing of topics to be addressed in a meeting.

Meeting notes   A written record of meeting attendees, topics addressed, decisions made, open issues, and responsibilities for open issues.

Meeting notice   A written announcement of a meeting. It generally contains the date, time, and location of the meeting, as well as the topics to be addressed.

Mock-ups   Stand-alone samples of completed work, such as a 6-foot-by-6-foot sample of a brick wall.

Network diagrams   Schedule that shows the relationships among the project activities with a series of nodes and connecting lines.

Notice to proceed   Written communication issued by the owner to the contractor, authorizing the contractor to proceed with the project and establishing the date for project commencement.

Occupational Safety and Health Administration   Federal agency responsible for establishing jobsite safety standards and enforcing them through inspection of construction worksites.

Officer in charge   General contractor's principal individual who supervises the project manager and is responsible for overall contract compliance.

Off-site construction   Prefabrication of building modules or systems, improving on-site cost and schedule performance.

Operation and maintenance manuals   A collection of descriptive data needed by the owner to operate and maintain equipment installed on a project.

Overbilling   Requesting payment for work that has not been completed.

Overhead   Expenses incurred that do not directly relate to a specific project – for example, rent on the contractor's home office.

Partnering   A cooperative approach to project management that recognizes the importance of all members of the project team, establishes harmonious working relationships among team members, and resolves issues in a timely manner.

Payment bond   See *labor and material payment bond*.

Payment request   Document or package of documents requesting progress payments for work performed during the period covered by the request, usually monthly.

Performance bond   A surety instrument that guarantees that the contractor will complete the project in accordance with the contract. It protects the owner from the general contractor's default and the general contractor from the subcontractor's default. A commonly used form is AIA document A312.

Performance standards   Standards a person is expected to achieve in the performance of his or her job.

Plugs   General contractor's cost estimates for subcontracted scopes of work.

Post-project analysis   Reviewing all aspects of the completed project to determine lessons that can be applied to future projects.

Pre-bid conference   Meeting of bidding contractors with the project owner and architect. The purpose of the meeting is to explain the project and bid process and solicit questions regarding the design or contract requirements.

Precedence-diagramming method   Scheduling technique that uses nodes to depict activities and arrows to depict relationships among the activities. Used by most scheduling software.

Preconstruction agreement   A short contract that describes the contractor's responsibilities and compensation for preconstruction services.

Preconstruction conference   Meeting conducted by owner or designer to introduce project participants and to discuss project issues and management procedures.

Preconstruction services   Services that a construction contractor performs for a project owner during design development and before construction starts.

Pre-final inspection   An inspection conducted when the project is near completion to identify all work that needs to be completed or corrected before the project can be considered completed.

Preparatory inspection   A quality control inspection to ensure that all preliminary work has been completed on a project site before starting the next phase of work.

Pre-proposal conference   Meeting of potential contractors with the project owner and architect. The purpose of the meeting is to explain the project, the negotiating process and selection criteria, and solicit questions regarding the design or contract requirements.

Prequalification of contractors   Investigating and evaluating prospective contractors based on selected criteria prior to inviting them to submit bids or proposals.

Product data sheet   Information furnished by a manufacturer to illustrate a material, product, or system for some portion of the project, which includes illustrations, standard schedules, performance data, instructions, and warranty; also known as material data or cut sheets.

Profit   The contractor's net income after all expenses have been subtracted.

Progress payments   Periodic (usually monthly) payments made during the course of a construction project to cover the value of work satisfactorily completed during the previous period.

Project close-out   Completing the physical construction of the project, submitting all required documentation to the owner, and financially closing out the project.

Project control   The methods a project manager uses to anticipate, monitor, and adjust to risks and trends in controlling costs, schedules, quality, and safety.

Project engineer   Project management team member who assists the project manager on larger projects. More experienced and has more responsibilities than the field engineer, but less than the project manager. Responsible for management of technical issues on the jobsite.

Project labor curve   A plot of estimated labor hours or crew size required per month for the duration of the project.

Project management   Application of knowledge, skills, tools, and techniques to the many activities necessary to complete a project successfully.

Project manager   The leader of the contractor's project team who is responsible for ensuring that all contract requirements are achieved safely and within the desired budget and time frame.

Project manual   A volume usually containing the instructions to bidders, the bid form, general conditions, and special conditions. It also may include a *geotechnical report*.

Project planning   The process of selecting the construction methods and the sequence of work to be used on a project.

Project-specific safety plan   A detailed accident prevention plan that is focused directly on the hazards that will exist on a specific project and on measures that can be taken to reduce the likelihood of accidents.

Project start-up   Mobilizing the project management team, establishing the project management office, and creating project document management systems.

Project team   Individuals from one or several organizations who work together as a cohesive team to construct a project.

Public–private partnership   Public agency partners with a contractor or developer – in the case of construction, to reduce costs and lawsuits, and ultimately save the taxpayer.

Pull planning   Scheduling method often utilizing stickie notes where milestones of each design or construction discipline are established and the project is scheduled backwards with the aid of short-term detailed schedules.

Punch list   A list of items that need to be corrected or completed before the project can be considered completed.

Purchase orders   Written contracts for the purchase of materials and equipment from suppliers.

Quality control   Process to assure materials and installations meet or exceed the requirements of the contract documents.

Quantity take-off   One of the first steps in the estimating process to measure and count items of work to which unit prices will later be applied to determine a project cost estimate.

Reimbursable costs   Costs incurred on a project that are reimbursed by the owner. The categories of costs that are reimbursable are specifically stated in the contract agreement.

Request for information   Document used to clarify discrepancies between differing contract documents and between assumed and actual field conditions; also known as *field question*.

Request for information log   Spreadsheet for tracking RFIs from initiation through designer response.

Request for proposal Document containing instructions to prospective contractors regarding documentation required and the process to be used in selecting the contractor for a project.

Request for qualifications A request for prospective contractors or subcontractors to submit a specific set of documents to demonstrate the firm's qualifications for a specific project.

Request for quotation A request for a prospective subcontractor to submit a quotation for a defined scope of work.

Retention A portion withheld from progress payments for contractors and subcontractors to create an account for finishing the work of any parties not able to or unwilling to do so; also known as retainage.

Rough order of magnitude cost estimate A conceptual cost estimate usually based on the size of the project. It is prepared early in the estimating process to establish a preliminary budget and decide whether or not to pursue the project.

Schedule of submittals A listing of all submittals required by the contract specifications.

Schedule of values An allocation of the entire project cost to each of the various work packages required to complete a project; used to develop a cash flow curve for an owner and to support requests for progress payments. Serves as the basis for AIA document G703, which is used to justify *progress payments.*

Schedule update Schedule revision to reflect the actual time spent on each activity to date.

Self-performed work Project work performed by the general contractor's workforce rather than by a subcontractor.

Shop drawing Drawing prepared by a contractor, subcontractor, vendor, or manufacturer to illustrate construction materials, dimensions, installation, or other information relating to the incorporation of the items into a construction project.

Short-form purchase order Purchase orders used on project sites by superintendents to order materials from local suppliers.

Short-interval schedule Schedule that lists the activities to be completed during a short interval (two to four weeks). Used by the superintendent and foremen to manage the work; also known as look-ahead schedule.

Site logistics plan Pre-project planning tool created often by the general contractor's superintendent that incorporates several elements, including temporary storm water control, hoisting locations, parking, trailer locations, fences, traffic plans, etc.

Soils report See *geotechnical report.*

Special conditions A part of the construction contract that supplements and may also modify, add to, or delete portions of the general conditions; also known as supplementary conditions.

Specialty contractors Construction firms that specialize in specific areas of construction work, such as painting, roofing, or mechanical. Such firms typically are involved in construction projects as subcontractors.

Start-up log   A spreadsheet used to manage start-up activities relating to suppliers and subcontractors.

Subcontractor call sheet   A form used to list all of the bidding firms from which the general contractor is soliciting subcontractor and vendor quotations.

Subcontractor preconstruction meeting   A meeting the project manager and/ or superintendent conduct with each subcontractor before allowing him or her to start work on a project.

Subcontractors   Specialty contractors who contract with and are under the supervision of the general contractor.

Subcontracts   Written contracts between the general contractor and specialty contractors who provide craft labor and usually material for specialized areas of work.

Submittals   Shop drawings, product data sheets, and samples submitted by contractors and subcontractors for verification by the design team that the materials purchased for installation comply with the design intent.

Substantial completion   State of a project when it is sufficiently completed that the owner can use it for its intended purpose.

Summary schedule   Abbreviated version of a detailed construction schedule that may include 10 to 20 major activities.

Superintendent   Individual from the contractor's project team who is the leader on the jobsite and who is responsible for supervision of daily field operations on the project.

Surety   A company that provides a bond guaranteeing that another company will perform in accordance with the terms of a contract.

Sustainability   Broad term incorporating many green building design and construction goals and processes, including LEED.

Technical specifications   A part of the construction contract that provides the qualitative requirements for a project in terms of materials, equipment, and workmanship.

Telephone memorandum   A written summary of a telephone conversation.

Third-tier subcontractor   A subcontractor who is hired by a firm that has a subcontract with the general contractor.

Time and materials contract   A cost-plus contract in which the owner and the contractor agree to a labor rate that includes the contractor's profit and overhead. Reimbursement to the contractor is made based on the actual costs for materials and the agreed labor rate times the number of hours worked.

Total quality management   A management philosophy that focuses on continuous process improvement and customer satisfaction.

Traditional project delivery method   A delivery method in which the owner has a contract with an architect to prepare a design for a project. When the design is completed, the owner hires a contractor to construct the project.

Transmittal   A form used as a cover sheet for formally transmitting documents between parties.

Unconditional lien release   A lien release indicating that the issuer has received a certain amount of payment and releases all lien rights associated with that amount.

Unit price contract   A contract that contains an estimated quantity for each element of work and a unit price. The actual cost is determined once the work is completed and the total quantity of work measured.

Value engineering   A study of the relative value of various materials and construction techniques to identify the least costly alternative without sacrificing quality or performance.

Warranty   Guarantees that all materials furnished are new and able to perform as specified, and that all work is free from defects in material or workmanship.

Work breakdown structure   A list of significant work items that will have associated cost or schedule implications.

Workers' compensation insurance   Protects the contractor from a claim due to injury or death of an employee on the project site.

Work package   A defined segment of the work required to complete a project.